中国轻工业“十三五”规划立项教材

高分子材料与工程专业系列教材

INTRODUCTION TO POLYMER SCIENCE AND ENGINEERING

(Second Edition)

聚合物科学与工程导论

(第二版)

〔英汉双语〕

揣成智　主　编
万　同　副主编

中国轻工业出版社

图书在版编目（CIP）数据

聚合物科学与工程导论：英汉双语/揣成智主编．—2版．—北京：中国轻工业出版社，2021．11

书名原文：Introduction to Polymer Science and Engineering

中国轻工业“十三五”规划立项教材．高分子材料与工程专业系列教材

ISBN 978-7-5184-2476-4

Ⅰ．①聚… Ⅱ．①揣… Ⅲ．①聚合物-高等学校-教材-英、汉 Ⅳ．①O63

中国版本图书馆CIP数据核字（2019）第088136号

责任编辑：林 媛

策划编辑：林 媛　责任终审：滕炎福　封面设计：锋尚设计

版式设计：王超男　责任校对：晋 洁　责任监印：张 可

出版发行：中国轻工业出版社（北京东长安街6号，邮编：100740）

印　刷：北京君升印刷有限公司

经　销：各地新华书店

版　次：2021年11月第2版第2次印刷

开　本：787×1092　1/16　印张：17．5

字　数：448千字

书　号：ISBN 978-7-5184-2476-4　定价：52．00元

邮购电话：010-65241695

发行电话：010-85119835　传真：85113293

网　址：http://www.chlip.com.cn

Email：club@chlip.com.cn

如发现图书残缺请与我社邮购联系调换

211452J1C202ZBW

前　言

本教材是中国轻工业“十三五”规划立项教材，其前身是《高分子材料工程专业英语》，2010年改名为《聚合物科学与工程导论》，是应中国轻工业出版社委托，及国内二十几所轻工院校高分子材料与工程专业的要求而编写和修订的。自1999年该教材被列为本科生必修专业课程以来，该教材被国内20几所大学列为本科专业必修课程，经过近20年多届学生使用，效果较好。此次修订，除对第一版教材中各章内容，针对近年来的发展变化进行了修改和充实外，还增写了聚合物热分析单元。

本教材分为七个单元共45课，内容涉及聚合物发展史、聚合物科学基本概念、聚合物基础知识、聚合物合成、聚合物性能、聚合物热分析、聚合物材料和聚合物成型加工。

课文正文全部选用英文原版经典教材，使学生真正领会英文原版教材中专业知识的精髓。本教材既可作为高分子材料与工程专业必修的基础课，又可作为该领域专业课的双语教材或专业英语教学用书，也可作为从事高分子材料与工程研究领域的科技人员、教师及研究生提高业务及其专业英语水平的学习参考书。

本教材每课都附有练习题和阅读材料，便于学生对课文内容的理解，其阅读取材丰富，形式灵活，图文并茂、直观生动、深入浅出，简明易懂。每课除正文和阅读材料外，都标出了专业词汇的生词，并注有音标、常用短语和词组，并将课文的难点做了中文注释，可与中文内容对照学习。教材书后列出了总词汇表与术语的中英文对照表，便于查阅。

本教材由天津科技大学的揣成智、万同、褚立强、王彪编写。全书由主编揣成智、副主编万同统稿。

在本教材编写过程中，得到中国轻工业联合会、中国轻工业出版社、天津科技大学及兄弟院校有关领导和同仁的帮助与支持，谨此致谢。

由于水平所限，难免有不足和错漏之处，诚恳希望使用本书的读者批评指正。

编者

2019年3月

Content 目　录

PART 1 INTRODUCTION
第一部分 概　述

PART 2 FUNDAMENTALS OF POLYMER
第二部分 聚合物基础

PART 3 POLYMER SYNTHESIS
第三部分 聚合物合成

PART 4 POLYMER PROPERTIES
第四部分 聚合物性能

PART 5　THERMAL ANALYSIS OF POLYMERS
第五部分 聚合物热分析

PART 6　POLYMER MATERIALS
第六部分　聚合物材料

PART 7 POLYMER MOLDING AND PROCESSING
第七部分 聚合物成型加工

PART 1　INTRODUCTION

Lesson 1　Introduction to History of Polymer Science

Since the Second World War, polymeric materials have been the fastest-growing segments of the world chemical industry. It has been estimated that more than a third of the chemical research money is spent on polymers, with a correspondingly large proportion of technical personnel working in the area.

A modern automobile contains over 150 kg of plastics, and this does not include paints, the rubber in tires, or the fibers in tires and upholstery. New aircraft incorporate increasing amounts of polymers and polymer-based composites. With the need to save fuel and therefore weight, polymers will continue to replace traditional materials in the automotive and aircraft industries. Similarly, the applications of polymers in the building construction industry (piping, resilient flooring, siding, thermal and electrical insulation, paints, decorative laminates) are already impressive, and will become even more so in the future. A trip through a supermarket will quickly convince anyone of the importance of polymers in the packaging industry (bottles, films, trays). Many other examples could be cited, but to make a long story short, the use of polymers now outstrips that of metals on a mass basis.

People have objected to synthetic polymers because they are not "natural." Well, botulism is natural, but it's not particularly desirable. Seriously, if all the polyester and nylon fibers in use today were to be replaced by cotton and wool, their closest natural counterparts, calculations show that there wouldn't be enough arable land left to feed the populace[①] and we'd be overrun by sheep. The fact is there simply are no practical natural substitutes for many of the synthetic polymers used in modern society.

Since nearly all modern polymers have their origins in petroleum, it has been argued that this increased reliance on polymers constitutes an unnecessary drain on energy resources. However, the raw materials for polymers account for less than two percent of total petroleum and natural gas consumption, so even the total elimination of synthetic polymers would not contribute significantly to the conservation of hydrocarbon resources. Furthermore, when total energy costs (raw materials plus energy to manufacture and ship) are compared, the polymeric item often comes out well ahead of its traditional counterpart, e. g., glass vs. plastic beverage bottles. In addition, the manufacturing processes used to produce polymers often generate considerably less environmental pollution than the processes used to produce the traditional counterparts, e. g., polyethylene film vs. kraft paper for packaging.

Ironically, one of the most valuable properties of polymers, their chemical inertness, causes problems because polymers do not normally degrade in the environment. As a result, they contribute increasingly to litter and the consumption of scarce landfill space②. Progress is being made toward the solution of these problems. Environmentally degradable polymers are being developed, although this is basically a wasteful approach and we're not yet sure of the impact of the degradation products. Burning polymer waste for its fuel value makes more sense, because the polymers retain essentially the same heating value as the raw hydrocarbons from which they were made. Still, the polymers must be collected and this approach wastes the value added in manufacturing the polymers.

The ultimate solution is recycling. If waste polymers are to be recycled, they must first be collected. Unfortunately, there are literally dozens (maybe hundreds) of different polymers in the waste mix, and mixed polymers have mechanical properties about like cheddar cheese. Thus, for anything but the least-demanding applications (e. g., parking bumpers, flower pots), the waste mix must be separated prior to recycling. To this end, automobile manufacturers are attempting to standardize on a few well-characterized plastics that can be recovered and re-used when the car is scrapped. Many objects made of the large-volume commodity plastics now have molded-in identifying marks, allowing hand sorting of the different materials.

Processes have been developed to separate the mixed plastics in the waste. The simplest of these is a sink-float scheme which takes advantage of density differences among the various plastics. Unfortunately, many plastic items are foamed, plated, or filled (mixed with nonpolymer components), which complicates density-based separations. Other separation processes are based on solubility differences between various polymers. An intermediate approach chemically degrades the waste polymer to the starting materials from which new polymer can be made.

1. New words

paint[peint] *n.* 颜料,油漆

rubber['rʌbə] *n.* 橡胶

upholstery[ʌp'həulstəri] *n.* 室内装饰品

aircraft['ɛəkraːft] *n.* 航空器

siding['saidiŋ] *n.* 板壁

polyester[ˌpɔli'estə] *n.* 聚酯

nylon['nailən] *n.* 尼龙

botulism['bɔtjulizəm] *n.* 肉毒中毒(食物中毒的一种)

counterpart['kauntəpaːt] *n.* 对应物

arable['ærəbl] *adj.* 可耕的,可开垦的

overrun[ˌəuvə'rʌn] *n.* 超出限度;*vt.* & *vi.* 泛滥

petroleum[pi'trəuliəm] *n.* 石油

hydrocarbon['haidrəu'kaːbən] *n.* 烃,碳氢化合物

beverage['bevəridʒ] *n.* 饮料

polyethylene[ˌpɔli'eθiliːn] *n.* 聚乙烯

degrade[di'greid] *vt.* & *vi.* (使)降解,(使)退化
scarce[skεəs] *adj.* 缺乏的,不足的
landfill['lændfil] *n.* 垃圾,垃圾掩埋法
bumper['bʌmpə] *n.* 缓冲器
component[kəm'pəunənt] *adj.* 组成的,合成的,成分的,分量的
cheddar['tʃedə] *n.* 干酪的一种

2. Phrases and expressions

polymer-based composites 聚合物基复合材料
resilient floor 弹性地板
electrical insulation 电绝缘
decorative laminates 装饰层压板
packaging industry 包装工业
energy resource 能源
raw material 原材料
account for 占,说明
kraft paper 牛皮纸

3. Notes to the text

①People have objected to synthetic polymers because they're not "natural." Well, botulism is natural, but it's not particularly desirable. Seriously, if all the polyester and nylon fibers in use today were to be replaced by cotton and wool, their closest natural counterparts, calculations show that there wouldn't be enough arable land left to feed the populace,…… 人们反感合成聚合物是因为合成聚合物不是"天然产物"。然而,食物中毒也是一种自然现象,但它却不是人们所需要的。严格说来,如果现今使用的聚酯和尼龙纤维都由棉花和羊毛这些自然界中最相近的纤维来替代,则有计算表明留给人类可耕种的土地将满足不了人们的粮食需要了。

②As a result, they contribute increasingly to litter and the consumption of scarce landfill space. 结果,垃圾日益增加,垃圾场日渐减少。

4. Exercises

(1) Consider the room you're in.

a. Identify the items in it that are made of polymers.

b. What would you make those items of if there were no polymers?

c. Why do you suppose polymers were chosen over competing materials (if any) for each particular application?

(2) Repeat Problem 1 for your automobile. Don't forget to look under the hood.

(3) You wish to develop a polymer to replace glass in widow glazing. What properties must a polymer have for that application?

Reading Material

History of Macromolecular Science

Natural polymers have been utilized throughout the ages. Since his beginning man has been dependent upon animal and vegetable matter for sustenance, shelter, warmth, and other requirements and desires. Natural resins and gums have been used for thousands of years. Asphalt was utilized in pre-Biblical times; amber was known to the ancient Greeks; and gum mastic was used by the Romans.

In the search by the early organic chemists for pure compounds in high yields, many polymeric substances were discovered and as quickly discarded as oils, tars, or undistillable residues. A few of these materials, however, attracted interest. Poly(ethylene glycol) was prepared about 1860; the individual polymers with degree of polymerization up to 6 were isolated and their structures correctly assigned. The concept of extending the structure to very high molecular weights by continued condensation was understood.

Other condensation polymers were prepared in succeeding decades. As the molecular aggregation theories gained in popularity, structures involving small rings held together by secondary bond forces were often assigned to these products①.

Some vinyl polymers were also discovered. Styrene was polymerized as early as 1839, isoprene in 1879, and methacrylic acid in 1880. Again cyclic structures held together by "partial valences" were assigned.

Acceptance of the macromolecular hypothesis came about in the 1920's, largely because of the efforts of Staudinger, who received the Nobel Prize in 1953 for his championship of this viewpoint②. In 1920 he proposed long-chain formulas for polystyrene, rubber, and polyoxymethylene. His extensive investigations of the latter polymers left no doubt as to their long-chain nature. More careful molecular weight measurements substantiated Staudinger's conclusions, as did x-ray studies showing structures for cellulose and other polymers which were compatible with chain formulas. The outstanding series of investigations by Carothers beginning in 1929 supplied quantitative evidence substantiating the macromolecular viewpoint.

One deterrent to the acceptance of the macromolecular theory was the problem of the ends of the long-chain molecules. Since the degree of polymerization of a typical polymer is several hundred, chemical methods for detecting end groups were at first not successful. Staudinger suggested that no end groups were needed to saturate terminal valences of the long chains; they were considered to be unreactive because of the size of the molecules. Large ring structures were also hypothesized; and this concept was popular for many years. Not until Flory elucidated the mechanism for chain-reaction polymerization did it become clear that the ends of long-chain molecules consist of normal, satisfied valence structures③. The presence and nature of end groups have since been investigated in detail by chemical methods.

Staudinger was among the first to recognize the large size of polymer molecules, and to utilize the dependence on molecular weight of a physical property, such as dilute solution viscosity, for determining polymer molecular weights. He also understood clearly that synthetic polymers are polydisperse. A few years later, Lansing and Kraemer distinguished unmistakably among the various average molecular weights obtainable experimentally.

Staudinger's name is also associated with the first studies of the configuration of polymer chain atoms. He showed that the phenyl groups in polystyrene are attached to alternate chain carbon atoms. This regular head-to-tail configuration has since been established for most vinyl polymers. The mechanism for producing branches in normally linear vinyl polymers was introduced by Flory but such branches were not adequately identified and characterized for another decade. Natta first recognized the presence of stereospecific regularity in vinyl polymers.

1. New words

gum[gʌm]*n.* 胶
asphalt['æsfælt]*n.* 沥青,柏油
amber['æmbə]*n.* 琥珀
tar[taː]*n.* 焦油
vinyl['vainil]*n.* 乙烯基
styrene['staiəriːn]*n.* 苯乙烯
glycol['glaikəl]*n.* 乙二醇
isoprene[ai'səupriː]*n.* 异戊二烯
methacrylic[me'θəkrilik]*adj.* 甲基丙烯类的
polystyrene[ˌpɔli'staiəriːn]*n.* 聚苯乙烯
polyoxymethylene[ˌpɔliˌɔksi'meθiliːn]*n.* 聚甲醛
cellulose['seljuləus]*n.* 纤维素
deterrent[di'terənt]*n.* 阻碍物
viscosity[vis'kɔsiti]*n.* 黏度
distillable[dis'tiləbl]*adj.* 可由蒸馏而得的
residuum[ri'zidjuəm]*n.* 剩余,残滓
polydisperse[ˌpɔlidis'pəːs]*adj.* 多分散性的
phenyl['fenil]*n.* 苯基
configuration[kənˌfigju'reiʃən]*n.* 构型
stereospecific[ˌstiəriəuspə'sifik]*adj.* 有规立构定向的
elucidate[i'ljuːsideit]*vt.* 阐明

2. Phrases and expressions

ethylene glycol 乙二醇
degree of polymerization 聚合度
molecular weight 分子量
secondary bond 次合键,次价力键
methacrylic acid 甲基丙烯酸,丙烯酸
long-chain 长链
end group 端基
chain-reaction 链式反应
dilute solution viscosity 稀溶液黏度
head-to-tail configuration 头尾构型

3. Notes to the text

①As the molecular aggregation theories gained in popularity, structures involving small rings held together by secondary bond forces were often assigned to these products. 由于分子缔合理论当时很盛行,人们经常把这些产品的结构看成是由次合键力结合起来的小环。

②Acceptance of the macromolecular hypothesis came about in the 1920's, largely because of the efforts of Staudinger, who received the Nobel Prize in 1953 for his championship of this viewpoint.　主要由于 Staudinger 的努力,20 世纪 20 年代高分子假说才被人们接受。由于他对这个观点的大力支持和提倡,1953 年他被授予诺贝尔奖。

③Not until Flory elucidated the mechanism for chain-reaction polymerization did it become clear that the ends of long-chain molecules consist of normal, satisfied valence structures.　直到 Flory 澄清链式反应聚合机理以后,人们才弄清楚长链分子反应末端是正常的价键结构。

Lesson 2 Basic Concepts of Polymer Science

Almost half a century ago, Wolfgang Ostwald coined the phrase "the land of neglected dimensions" to describe the range of sizes between molecular and macroscopic within which occur most colloidal particles. The term "neglected dimensions" might have been applied equally well to the world of polymer molecules, the high-molecular-weight compounds so important to man and his modern technology. It was not until the third decade of this century that the science of high polymers began to emerge, and the major growth of the technology of these materials came even later. Yet today polymer dimensions are neglected no more, for industries associated with polymeric materials employ more than a third of all American chemists and chemical engineers.

The science of macromolecules is divided between biological and nonbiological materials. Each is of great importance. Biological polymers form the very foundation of life and intelligence, and provide much of the food on which man exists. This book, however, is concerned with the chemistry, physics, and technology of nonbiological polymers. These are primarily the synthetic materials used for plastics, fibers, and elastomers, but a few naturally occurring polymers, such as rubber, wool, and cellulose, are included. Today, these substances are truly indispensable to mankind, being essential to his clothing, shelter, transportation, and communication, as well as to the conveniences of modern living.

A polymer is a large molecule built up by the repetition of small, simple chemical units. In some cases the repetition is linear, much as a chain is built up from its links. In other cases the chains are branched or interconnected to form three-dimensional networks①. The repeat unit of the polymer is usually equivalent or nearly equivalent to the monomer, or starting material from which the polymer is formed. Thus the repeat unit of poly (vinyl chloride) is $—CH_2CHCl—$; its monomer is vinyl chloride, $CH_2=CHCl$.

The length of the polymer chain is specified by the number of repeat units in the chain. This is called the degree of polymerization. The molecular weight of the polymer is the product of the molecular weight of the repeat unit and the degree of polymerization②. Using poly (vinyl chloride) as an example, a polymer of degree of polymerization 1,000 has a molecular weight of $63 \times 1,000 = 63,000$. Most high polymers useful for plastics, rubbers, or fibers have molecular weights between 10,000 and 1,000,000.

Unlike many products whose structure and reactions were well known before their industrial application, some polymers were produced on an industrial scale long before their chemistry or physics was studied. Empiricism in recipes, processes, and control tests was usual.

Gradually the study of polymer properties began. Almost all were first called anomalous because they were so different from the properties of low-molecular-weight compounds. It was soon realized, however, that polymer molecules are many times larger than those of ordinary substances. The presumably anomalous properties of polymers were shown to be normal for such materials, as the consequences of their size were included in the theoretical treatments of their properties.

Primary chemical bonds along polymer chains are entirely satisfied. The only forces between molecules are secondary bond forces of attraction, which are weak relative to primary bond forces③. The high molecular weight of polymers allows these forces to build up enough to impart excellent strength, dimensional stability, and other mechanical properties to the substances④.

1. New words

colloidal[kə'lɔidl] *adj.* 胶状的,胶质的
coin[kɔin] *vt.* 创造
elastomer[i'læstəumə] *n.* 弹性体
monomer['mɔnəmə] *n.* 单体
recipe['resipi] *n.* 配方,处方
indispensable[indis'pensəbl] *adj.* 不可缺少的
empiricism[em'pirisizəm] *n.* 经验主义
anomalous[ə'nɔmələs] *adj.* 反常的,不规则的

2. Phrases and expression

build up　组成,形成,聚集
associate with　与……相关,联合
three dimensional network　三维网状结构
repeat unit　重复单元
poly(vinyl chloride)　聚氯乙烯
primary bond　主价键
dimensional stability　尺寸稳定性

3. Notes to the text

① In some cases the repetition is linear, much as a chain is built up from its links. In other cases the chains are branched or interconnected to form three-dimensional networks.　这些重复结构有的是线型的,很像由一个一个的环构成的一条链子;有时这些链会分叉或相互连接成三维网状结构。

② The molecular weight of the polymer is the product of the molecular weight of the repeat unit and the degree of polymerization.　聚合物的相对分子质量是重复单元的相对分子质量与聚合度的乘积。

③ The only forces between molecules are secondary bond forced of attraction, which are weak relative to primary bond forces.　分子间唯一能有的力是次价键引力,这种引力比主价键的力要小。

④ The high molecular weight of polymers allows these forces to build up enough to impart excellent strength, dimensional stability and other mechanical properties to the substances. 由于聚合物的相对分子质量大,这些力可以累加得很大,使这些物质具有非常好的强度、尺寸稳定性和其他机械性能。

4. Exercises

(1) What phrase did Wolfgang Ostwald coin to describe the macromolecules almost half a century ago?

(2) What is the science of macromolecules divided? What is this book concerned with?

(3) What is the definition of a polymer?

(4) Why were almost polymers, first called anomalous in the study of polymer properties?

(5) How much is its molecular weight if a polyethylene has the degree of polymerization 1,000?

Reading Material

Basic Concepts of Polymers

There are five major areas of application for polymers: (1) plastics, (2) rubbers or elastomers, (3) fibers, (4) surface finishes and protective coatings, and (5) adhesives. Despite the fact that all five applications are based on polymers, and in many cases the same polymer is used in two or more, the industries grew up pretty much separately. It was only after Dr. Herman Staudinger proposed the "macromolecular hypothesis" in the 1920s explaining the common molecular makeup of these materials (for which he won the 1953 Nobel Prize in chemistry in belated recognition of the importance of his work) that polymer science began to evolve from the independent technologies. Thus, a sound fundamental basis was established for continued technological advances. The history of polymer science is treated in detail elsewhere.

Economic considerations alone would be sufficient to justify the impressive scientific and technological efforts expended on polymers in the past several decades. In addition, however, this class of materials possesses many interesting and useful properties that are completely different from those of the more traditional engineering materials, and that cannot be explained or handled in design situations by the traditional approaches. A description of three simple experiments should make this obvious.

Silly putty, a silicone polymer, bounces like rubber when rolled into a ball and dropped. On the other hand, if the ball is placed on a table, it will gradually spread to a puddle. The material behaves as an elastic solid under certain conditions, and as a viscous liquid under others.

If a weight is suspended from a rubber band, and the band is then heated (taking care not to burn it), the rubber band will contract appreciably. All materials other than polymers will undergo the expected thermal expansion upon heating (assuming that no phase transformation has occurred over the temperature range)[①].

When a rotating rod is immersed in a molten polymer or a fairly concentrated polymer solution, the liquid will actually climb up the rod. This phenomenon, the Weissenberg effect, is contrary to what is observed with nonpolymer liquids, which develop a curved surface profile with a lowest point at the rod, as the material is flung outward by centrifugal force[②].

Although such behaviour is unusual in terms of the more familiar materials, it is a perfectly logical consequence of the *molecular structure* of polymers. This molecular structure is the key to an understanding of the science and technology of polymers, and will underlie the chapters to follow.

Figure 2.1 illustrates the questions to be considered:

1. How is the desired molecular structure obtained?
2. How do the polymer's processing (i.e., formability) properties depend on its molecular structure?
3. How do its material properties (mechanical, chemical, optical, etc.) depend on molecular structure?
4. How do material properties depend on a polymer's processing history?
5. How do its applications depend on its material properties?

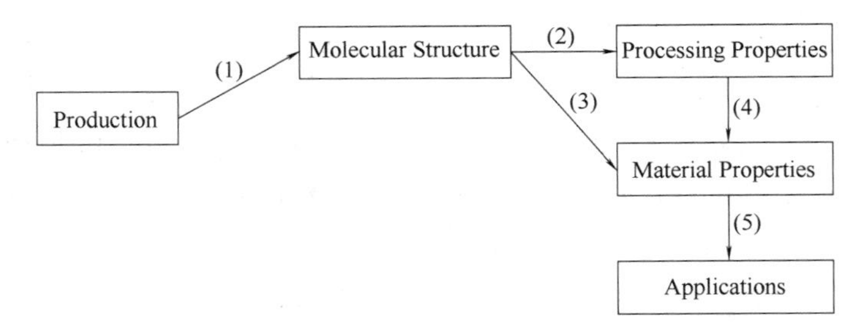

Figure 2.1 The key role of molecular structure in polymer science and technology

The word polymer comes from the Greek for "many-membered." Strictly speaking, it could be applied to any large molecule that is formed from a relatively large number of smaller units or "mers," for example, a sodium chloride crystal, but it is most commonly (and exclusively, here) restricted to materials in which mers are held together by covalent bonding, that is, shared electrons[③]. For our purposes, only a few bond valences need be remembered:

$$-\overset{|}{\underset{|}{C}}- \quad \overset{|}{N}\diagup\diagdown \quad -O- \quad Cl- \quad F- \quad H- \quad -\overset{|}{\underset{|}{Si}}-$$

It is always a good idea to "count the bonds" in any structure written to make sure they conform to the above.

Keep in mind also that when the covalently bound atoms differ in electronegativity, the electrons are not shared evenly between them. In gaseous HCl, for example, the electrons cluster around the electronegative chlorine atom, giving rise to a molecular dipole④:

$$\underset{\delta^{+}}{H} \quad \underset{\delta^{-}}{:\ddot{\underset{..}{Cl}}:}$$

Electrostatic forces between such dipoles can play an important role in determining polymer properties.

A brief, concise review of organic chemistry from the polymer standpoint is available. The most important constituents of living organisms, cellulose and proteins, are naturally occurring polymers, but we will confine our attention largely to synthetic polymers or to important modifications of natural polymers.

1. New words

polymer['pɔlimə] *n.* 聚合物
plastic['plæstik] *n.* 塑料;*adj.* 塑性的
hypothesis[hai'pɔθisis] *n.* 假设
belated[bi'leitid] *adj.* 误期的,迟来的
fiber['faibə] *n.* 纤维
adhesive[əd'hi:siv] *n.* 黏合剂;*adj.* 有黏性的
makeup['meikʌp] *n.* 组成(接通,补给,修理)
evolve[i'vɔlv] *vt.* ;*vi.* . 进展,进化,展开
fundamental[ˌfʌndə'mentl] *n.* 基本原理;*adj.* 基本的,根本的
elsewhere['els 'hwɛə] *adv.* 在别处,到别处
putty['pʌti] *n.* 油灰,氧化锡;*v.* 用油灰填塞
bouncing['baunsiŋ] *adj.* 跳跃的,活泼的,巨大的
silicone['silikəun] *n.* 硅,硅有机树脂
gradually['grædjuəli] *adv.* 逐渐地
puddle['pʌdl] *n.* 胶泥;*vt.* 搅拌
rotate[rəu'teit] *vt.* ;*vi.* (使)旋转
rod[rɔd] *n.* 杆,棒
immerse[i'mə:s] *vt.* 浸,陷入
molten['məultən] *adj.* 熔化的,炽热的,铸造的
underlie[ˌʌndə'lai] *vt.* 位于…之下,成为…的基础
formability[fɔ:mə'biliti] *n.* 可成型性
mechanical[mi'kænikl] *adj.* 机械的,力学的
optical['ɔptikəl] *adj.* 眼睛的,视觉的,光学的
mer['mə:] *n.* 基体
exclusively[ik'sklu:sivli] *adv.* 仅仅
valence['veiləns] *n.* 化合价
electronegativity[i'lektrəuˌnegətiviti] *n.* 负电性

electron[i'lektrɔn]*n.* 电子
dipole['daipəul]*n.* 偶极
constituent[kən'stitjuənt]*n.* 成分;*adj.* 构成的,组织的
protein['prəuti:n]*n.* 蛋白质
modification[ˌmɔdifi'keiʃən] *n.* 修饰,改性

2. Phrases and expressions

surface finishes 表面加工,表面修饰
protective coating 保护涂层
macromolecular hypothesis 大分子假说
engineering material 工程材料
silly putty 弹性橡皮泥
rubber band 橡皮圈
thermal expansion 热膨胀
phase transformation 相转变
fairly concentrated polymer solution 聚合物浓溶液
Weissenberg effect 韦森堡效应
centrifugal force 地心引力
sodium chloride crystal 氯化钠晶体
covalent bonding 共价键
shared electrons 共价电子,共用电子对
electrons cluster 电子簇,电子雾,电子云
electronegative chlorine atom 负电性的氯原子
molecular dipole 分子偶极子
living organism 生命有机体
natural polymer 天然高分子

3. Notes to the text

① If a weight is suspended from a rubber band, and the band is then heated(taking care not to burn it), the rubber band will contract appreciably. All materials other than polymers will undergo the expected thermal expansion upon heating(assuming that no phase transformation has occurred over the temperature range). 如果给一个橡皮圈悬挂一个重力,然后加热(注意避免将其点燃),我们会发现橡皮圈将出现明显的收缩。除了聚合物以外的所有其他材料,在受热时都将经历一个热膨胀过程(假定在该温度范围内不发生相转变)。

② When a rotating rod is immersed in a molten polymer or a fairly concentrated polymer solution, the liquid will actually climb up the rod. This phenomenon, the Weissenberg effect, is contrary to what is observed with nonpolymer liquids, which develop a curved surface profile with a lowest point at the rod, as the material is flung outward by centrifugal force. 当一根旋转的杆浸入聚合物熔体或聚合物浓溶液时,液态物质将沿杆爬升。这种现象称为韦森堡效应。而在小分子液体中,物质由于离心力的作用将被向外抛出,形成杆所在位置低于液面其他位置的表面轮廓,该种现象刚好与韦森堡效应相反。

③ The word polymer comes from the Greek for "many-membered." Strictly speaking, it could be applied to any large molecule that is formed from a relatively large number of smaller units or "mers", for example, a sodium chloride crystal, but it is most commonly(and exclusively, here) restricted to materials in witch mers are held together by covalent bonding, that is, shared electrons. 聚合物一词源于希腊语,意为"多组分

的”。严格地讲,它可以指任何由相对来说大量的、小的单元或“基体”所组成的大分子,例如氯化钠单晶,但一般(仅)限于基体之间由共价键相连的材料,即基体之间通过共用电子对的方式相连的材料。

④ Keep in mind also that when the covalently bound atoms differ in electronegativity, the electrons are not shared evenly between them. In gaseous HCl, for example, the electrons cluster around the electronegative chlorine atom, giving rise to a molecular dipole... 同时需要记住的是,当由共价键约束的一对原子的负电性不同时,电子对在它们之间的分布是不平衡的。举例来说,在气态的氯化氢中,电子云分布在负电性较强的氯原子周围,使分子偶极增大……。

PART 2 FUNDAMENTALS OF POLYMER

Lesson 3 Types of Polymers

The large number of natural and synthetic polymers has been classified in several ways. These will be outlined below.

The earliest distinction between types of polymers was made long before any concrete knowledge of their molecular structure. It was a purely phenomenological distinction, based on their reaction to heating and cooling①.

Thermoplastics

It was noted that certain polymers would soften upon heating, and could then be made to flow when a stress was applied. When cooled again, they would reversibly regain their solid or rubbery nature. These polymers are known as *thermoplastics*. By analogy, ice and solder, though not polymers, behave as thermoplastics.

Thermoplastic polymers are characterized by softening upon heating and hardening by cooling. Since the giant molecules of these materials have no strong bonds between the individual molecules, they can be softened by heat and remolded over and over again. This is an advantage in molding processes such as extrusion or injection where scrap or rejected products can be reground and molded again. Some of the thermoplastic materials will burn freely when exposed to an open flame while others of this group will not support combustion.

Thermosets

Other polymers, although they might be heated to the point where they would soften and could be made to flow under stress *once*, would not do so reversibly; that is, heating caused them to undergo a "curing" reaction. Sometimes these materials emerge from the synthesis reaction in a cured state. Further heating of these *thermosetting* polymers ultimately leads only to degradation (as is sometimes attested to by the smell of a short-circuited electrical appliance) and not softening and flow②. Again by analogy, eggs and concrete behave as thermosets. Continued heating of thermoplastics will also lead ultimately to degradation, but they will generally soften at temperatures below their degradation point.

The group of thermosetting polymers, numbering less than the thermoplastic group, possesses quite different characteristics. Because of the irreversible reaction by which they polymerize, they form a rigid, hard and often brittle, infusible mass. The crosslinking molecular structure with strong chemical bonds between the polymer chains causes these materials to be rigid and hard as no slippage can occur between the polymer chains. Since all the bonds are strong, when the material is heated no chain flow or softening can occur. Intensive heating of a thermoset will cause breakage of the chemical bonds resulting in a charring of the material[3]. They are not flammable. In general, thermosetting plastics can be described as being hard, strong, and rigid, with good heat resistance.

Natural rubber is a classic example of these two categories. Introduced to Europe by Columbus, natural rubber did not achieve commercial significance for centuries; because it was a thermoplastic, articles made of it would become soft and sticky on hot days. In 1839, Charles Goodyear discovered the curing reaction with sulfur (which he called *vulcanization* in honor of the Roman god of fire) that converted the polymer to a thermoset[4]. This allowed the rubber to maintain its useful properties to much higher temperatures, which ultimately led to its great commercial importance.

1. New words

classify[ˈklæsifai] *vt.* 分类，归类

outline[ˈəutlain] *n.* 大纲；*vt.* 概要，描述要点

distinction[disˈtiŋkʃən] *n.* 差别，对比，区分

thermoplastics[ˌθəːməˈplæstiks] *n.* 热塑性塑料；*adj.* 热塑性的

reversibly[riˈvəːsəbli] *adv.* 可逆地

regain[riˈgein] *vt.* 恢复，重回

rubbery[ˈrʌbəri] *adj.* 强韧的

characterize[ˈkæriktəraiz] *vt.* 表示…的特色，赋予…的特色

remold[ˈriːˈməuld] *vt.* 改造，重铸

extrusion[eksˈtruːʒən] *n.* 挤出

injection[inˈdʒekʃən] *n.* 注射

scrap[skræp] *n.* 残余物；*vt.* 扔弃

thermosets[ˈθəːməsets] *n.* 热固性材料

emerge[iˈməːdʒ] *vi.* 浮现，形成

synthesis[ˈsinθisis] *n.* 合成，综合

ultimately[ˈʌltimətli] *adv.* 最后，最终

degradation[ˌdegrəˈdeiʃən] *n.* 退化，降解

concrete[ˈkɔnkriːt] *n.* 水泥，混凝土；*adj.* 具体的，实在的；*vi.* 凝结，结合

polymerize[ˈpɔliməraiz] *vi.* （使）聚合

rigid[ˈridʒid] *adj.* 僵硬的，刻板的，严格的

brittle[ˈbritl] *adj.* 易碎的

infusible[inˈfjuːzəbl] *adj.* 不溶解的

crosslinking[krɔsˈliŋkiŋ] *n.* 交联

slippage[ˈslipidʒ] *n.* 滑移，滑动

irreversible[ˌiriˈvəːsəbl] *adj.* 不可逆的，不能撤回的，不能取消的

intensive[inˈtensiv] *adj.* 集中的，强化的，深入的

breakage[ˈbreikidʒ] *n.* 裂口

charring[tʃɑːriŋ] *n.* 碳化，烧焦

flammable[ˈflæməbl] *adj.* 易燃的，可燃性的

category['kætigəri] *n.* 种类,类别
sticky['stiki] *adj.* 黏的,黏性的
sulfur['sʌlfə] *n.* 硫
vulcanization[ˌvʌlkənai'zeiʃən] *n.* 硫化
convert[kən'vəːt] *vt.* 使转变

2. Phrases and expressions

by analogy 用类比的方法,同样
thermoplastic polymer 热塑性聚合物
giant molecule 大分子
rejected product 次品,废品
irreversible reaction 不可逆反应
polymer chain 聚合物链
chemical bond 化学键
thermosetting plastics 热固性塑料
heat resistance 耐热性
natural rubber 天然橡胶

3. Notes to the text

① The earliest distinction between types of polymers was made long before any concrete knowledge of their molecular structure. It was a purely phenomenological distinction, based on their reaction to heating and cooling. 对不同种类聚合物最早的分类要远远早于人们对聚合物分子结构的任何具体认识。这种分类是基于聚合物对于冷和热的反应所作出的,是一种纯粹的现象层面上的区分。

② Further heating of these thermosetting polymers ultimately leads only to degradation (as is sometimes attested to by the smell of a short-circuited electrical appliance) and not softening and flow. 对热固性聚合物的进一步加热最终将导致其分解(有时可以通过闻到类似短路的电器所发出的气味来验证)而不会出现软化和流动。

③ Intensive heating of a thermoset will cause breakage of the chemical bonds resulting in a charring of the material. 对热固性材料加强热将会引起其化学键的断裂,导致材料碳化。

④ In 1839, Charles Goodyear discovered the curing reaction with sulfur (which he called vulcanization in honor of the Roman god of fire) that converted the polymer to a thermoset. 1839年,查尔斯·固特异发现了用硫作为固化剂的固化反应(为纪念罗马火神,他将其命名为硫化反应),硫能使聚合物变成一种热固性材料。

4. Exercises

(1) Try to expound the main distinctions between thermoplastics and thermosets according to the text. Their reaction to heating and cooling and different properties should be mainly considered.

(2) Classify the materials listed below as either thermoplastics or thermosets, then learn their applications through the Internet.

Polyethylene, phenolic resin (bakelite), polyvinyl chloride, epoxide resin, polystyrene

Reading Material

Chemistry of Synthesis

Pioneering workers in the field of polymer chemistry soon observed that they could produce polymers by two familiar types of organic reactions.

A. Condensation

Polymers formed from a typical organic condensation reaction, in which a small molecule (most often water) is split out, are known, logically enough, as *condensation polymers*①. The common esterification reaction of an organic acid and an organic base (alcohol) illustrates the simple "lasso chemistry" involved:

$$\underset{\text{Alcohol}}{R{-}O{-}H} + \underset{\text{Acid}}{HO{-}\overset{\overset{\displaystyle O}{\|}}{C}{-}R'} \longrightarrow \underset{\text{Ester}}{R{-}O{-}\overset{\overset{\displaystyle O}{\|}}{C}{-}R'} + H_2O$$

The —OH group on the alcohol and the $HO{-}\overset{\overset{\displaystyle O}{\|}}{C}{-}$ on the acid are known as *functional groups*, those parts of a molecule that participate in a reaction. Of course, the ester formed in the preceding reaction is not a polymer because we have only hooked up two small molecules, and the reaction is finished far short of anything that might be considered "many membered."

At this point it is useful to introduce the concept of functionality. Functionality is the number of bonds a mer can form with other mers in a reaction. In condensation polymerization, it is equal to the number of functional groups on the mer②.

B. Addition

The second polymer-formation reaction is known as addition polymerization and its products as addition polymers. Addition polymerizations have two distinct characteristics:

1. No molecule is split out; hence, the repeating unit has the same chemical formula as the monomer.
2. The polymerization reaction involves the opening of a double bond.

Monomers of the general type $\overset{|\quad|}{\underset{|\quad|}{C{=}C}}$ undergo addition polymerization:

$$x\overset{|\quad|}{\underset{|\quad|}{C{=}C}} \longrightarrow {-}\!\!\left[\overset{|\quad|}{\underset{|\quad|}{C{-}C}}\right]\!\!{-}_x$$

The double bond "opens up," forming bonds to other monomers at each end, so *a double bond* is *difunctional* according to our general definition of functionality.

An important subclass of the double-bond containing monomers is the vinyl monomers, $\begin{matrix} H & H \\ | & | \\ C & = C \\ | & | \\ H & X \end{matrix}$. Addition polymerization is occasionally referred to as vinyl polymerization.

1. New words

condensation [kɔnden'seiʃən] *n.* 压缩，缩聚

esterification [esˌterifi'keiʃən] *n.* 酯化作用

alcohol ['ælkəhɔl] *n.* 酒精，乙醇

ester ['estə] *n.* 酯，酯基

functionality [ˌfʌŋkəʃə'næliti] *n.* 官能度

distinct [dis'tiŋkt] *adj.* 不同的，明显的

difunctional [dai'fʌŋkʃənəl] *adj.* 双官能度的

subclass ['sʌbklɑːs] *n.* 子类，子集

vinyl ['vainil] *n.* 乙烯基

2. Phrases and expressions

organic reaction　有机反应

condensation reaction　缩合反应

split out　分裂，排出

condensation polymer　缩聚物

esterification reaction　酯化反应

organic acid　有机酸

organic base　有机碱

functional group　官能团

hook up　连接

addition polymerization　加成聚合反应

double bond　双键

3. Notes to the text

① Polymers formed from a typical organic condensation reaction, in which a small molecule (most often water) is split out, are known, logically enough, as condensation polymers.　典型的有机缩合反应会排出一个小分子（通常是水），由这种反应形成的聚合物也顺理成章地被称为缩聚物。

② Functionality is the number of bonds a mer can form with other mers in a reaction. In condensation polymerization, it is equal to the number of functional groups on the mer.　官能度是指一个基体分子在反应中与其他基体分子所能成键的数目。在缩聚反应中，官能度等于基体分子上官能团的数量。

Lesson 4 Configurational States

The term configuration refers to the "permanent" stereostructure of a polymer. The configuration is defined by the polymerization method, and a polymer preserves its configuration until it reacts chemically. A change in configuration requires the rupture of chemical bonds. Different configurations exist in polymers with stereocentres (tacticity) and double bonds (cis and trans forms). A polymer with the constitutional repeating unit —CH_2—CHX— exhibits two different stereoforms (configurational base units) for each constitutional repeating unit (Figure 4.1). The following convention is adapted to distinguish between the two stereoforms. One of the chain ends is first selected as the near one. The selected asymmetric carbon atom (the one with the attached X atom group of atoms) should be pointing upwards. The term d form is given to the arrangement with the X group pointing to the right (from the observer at the near end). The I form is the mirror-image of the d form, i. e. the X group points in this case to the left. This convention is, however, not absolute. If the near and far chain ends are reversed, i. e. if the chain is viewed from the opposite direction, the d and I notation is reversed for a given chain. When writing a single chain down on paper, the convention is that the atoms shown on the left-hand side are assumed to be nearer to the observer than the atoms on the right-hand side.

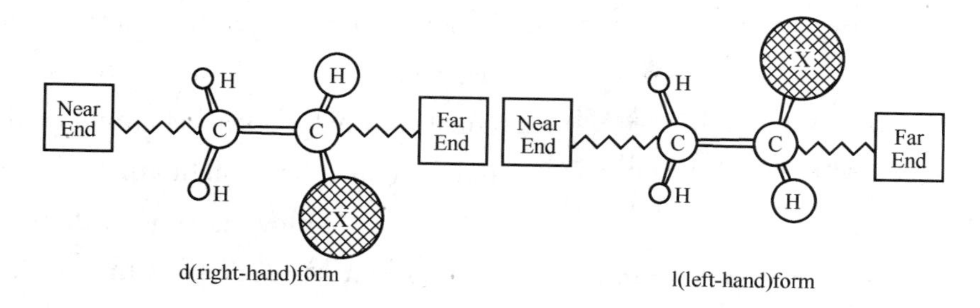

Figure 4. 1 Configurational base units of a polymer with the constitutional repeating unit —CH_2—CHX—. The 'X' is an atom or a group of atoms different from hydrogen

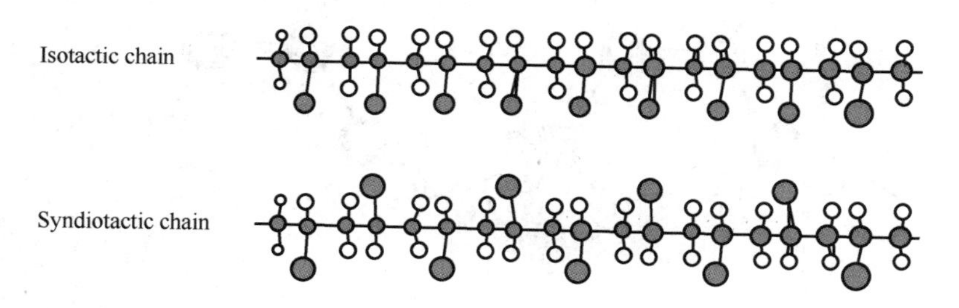

Figure 4. 2 Regular tactic chains of [—CH_2—CHX—]$_n$, where X is indicated by a filled circle

Tacticity is the orderliness of the succession of configurational base units in the main chain of a polymer molecule①. An isotactic polymer is a regular polymer consisting only of one species of configurational base unit, i. e. only the d or the I form (Figure 4. 2). These can be converted into each other by simple rotation of the whole molecule. Thus, in practice there is no difference between an all-d chain and an all-I chain. A small mismatch in the chain ends does not alter this fact. A syndiotactic polymer consists of an alternating sequence of the different configurational base units, i. e. . . . dldldldldldl. . . (Figure 4. 2). An atactic polymer has equal numbers of randomly distributed configurational base units.

Carbon-13 nuclear magnetic resonance (NMR) is the most useful method of assessing tacticity. By C-13 NMR it is possible to assess the different sequential distributions of adjacent configurational units that are called dyads, triads, tetrads and pentads. The two possible dyads are shown in Figure 4. 3. A chain with 100% meso dyads is perfectly isotactic whereas a chain with 100% racemic dyads is perfectly syndiotactic. A chain with a 50/50 distribution of meso and racemic dyads is atactic.

Triads express the sequential order of the configurational base units of a group of three adjacent constitutional repeating units②. The following triads are possible: mm, mr and rr (where meso = m; racemic = r). Tetrads include four repeating units and the following six sequences are possible: mmm, mmr, mrm, mrr, rmr and rrr. Sequences with a length of up to five constitutional repeating units, called pentads, can be distinguished by C-13 NMR. The following ten different pentads are possible: mmmm, mmmr, mmrm, mmrr, mrrm, rmmr, rmrm, mrrr, rmrr and rrrr.

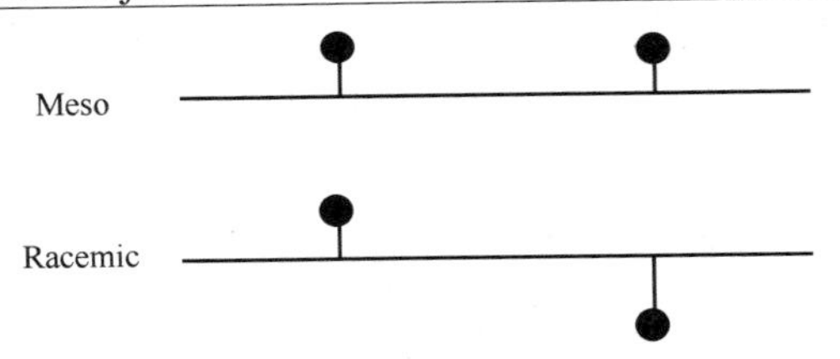

Figure 4. 3 Dyads of a vinyl polymer with the constitutional repeating unit —CH_2—CHX—, where the X group is indicated by a filled circle

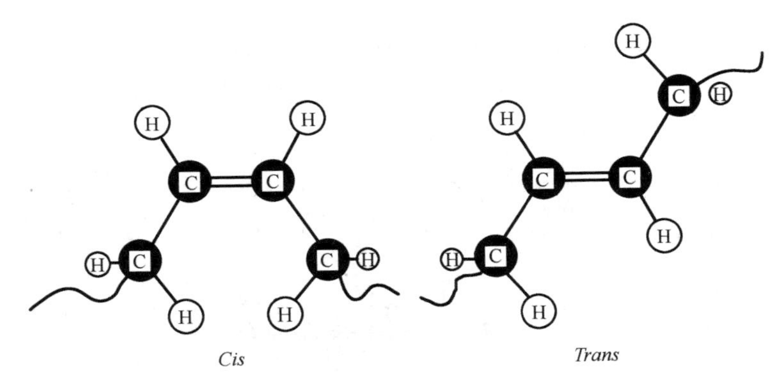

Figure 4. 4 Stereoforms of 1. 4-polybutadiene showing only the constitutional repeating unit with the rigid central double

Polymers with double bonds in the main chain, e. g. polydienes, show different stereostructures. Figure 4.4 shows the two stereoforms of 1, 4-polybutadiene: cis and trans. The double bond is rigid and allows no torsion, and the cis and trans forms are not transferable into each other. Polyisoprene is another well-known example: natural rubber consists almost exclusively of the cis form whereas gutta-percha is composed of the trans form. Both polymers are synthesized by 'nature', and this shows that stereoregularity was achieved in nature much earlier than the discovery of coordination polymerization by Ziegler and Natta.

The polymerization of diene monomers may involve different addition reactions (Figure 4.5). 1,2 addition yields a polymer with the double bond in the pendant group whereas 1,4 addition gives a polymer with the unsaturation in the main chain.

Vinyl polymers (—CH_2—CHX—) may show different configurations with respect to the head (CHX) and tail (CH_2): head-to-head, with—CHX bonded to CH_2—and head-to-head-tail-to-tail, with-CHX bonded to-CHX followed by—CH_2 bonded to —CH_2 (Figure 4.6).

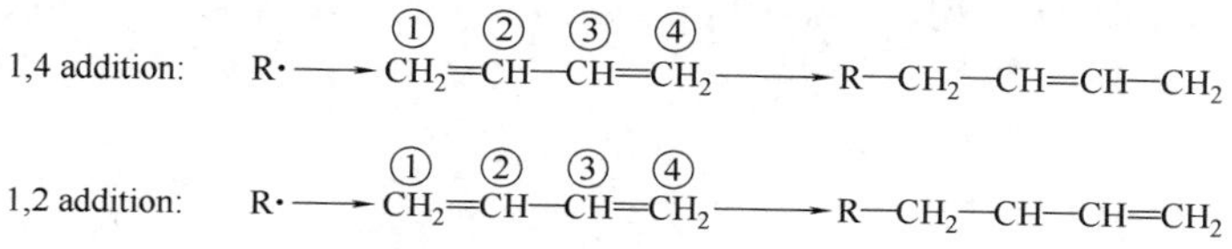

1,4 addition: R· ⟶ $\overset{①}{CH_2}$=$\overset{②}{CH}$—$\overset{③}{CH}$=$\overset{④}{CH_2}$ ⟶ R—CH_2—CH=CH—CH_2

1,2 addition: R· ⟶ $\overset{①}{CH_2}$=$\overset{②}{CH}$—$\overset{③}{CH}$=$\overset{④}{CH_2}$ ⟶ R—CH_2—CH—CH=CH_2

Figure 4.5 Different additions of butadiene.

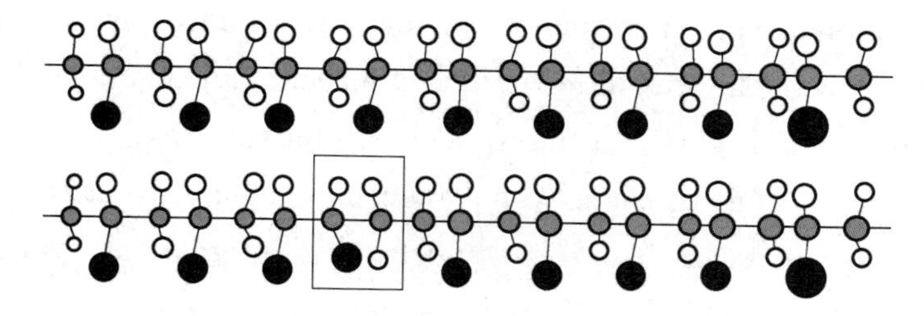

Figure 4.6 Head-to-tail configuration (upper chain) and a chain with a head-to-head junction followed by a tail-to-tail sequence (lower chain)

1. New words

configuration [kənˌfigjə'reʃən] *n.* 构型，配置，结构，外形

stereostructure ['stɛriostɛrIo] *n.* 立体结构

tacticity [tæk'tisəti] *n.* 立构规整度

syndiotadic [ˌsindaIo'tæktIk] *adj.* 间同的，间规的

stactic [e'tæktIk] *adj.* 不规则的，无规立构的

isotactic [ˌaisəu'tæktik] *adj.* 全同立构的，等规立构的

diene ['daii:n] *n.* 二烯（等于 diolefin）

conformation [ˌkɑnfɔr'meʃən] *n.* 构造，一致，符合，构象

2. Phrases and expressions

pointing upwards 指出向上
gutta-percha 杜仲胶
pendant group 侧基
coordination polymerization [高分子]配位聚合

3. Notes to the text

① Tacticity is the orderliness of the succession of configurational base units in the main chain of a polymer molecule. 立构规整度是聚合物分子主链中构型单元序列的有序性。

② Triads express the sequential order of the configurational base units of a group of three adjacent constitutional repeating units. 三元表示一组由相邻三个构成的重复单元的构型单元序列的顺序排列。

4. Exercise

(1) What is the definition of the configuration?

(2) Try to discuss the difference between configuration and conformation.

Reading Material

Conformational State

The configurational state determines the low temperature physical structure of the polymer. A polymer with an irregular configuration, e. g. an atactic polymer, will never crystallize and freezes to a glassy structure at low temperatures, whereas a polymer with a regular configuration, e. g. an isotactic polymer, may crystallize at some temperature to form a semicrystalline material. There is a spectrum of intermediate cases in which crystallization occurs but to a significantly reduced level.

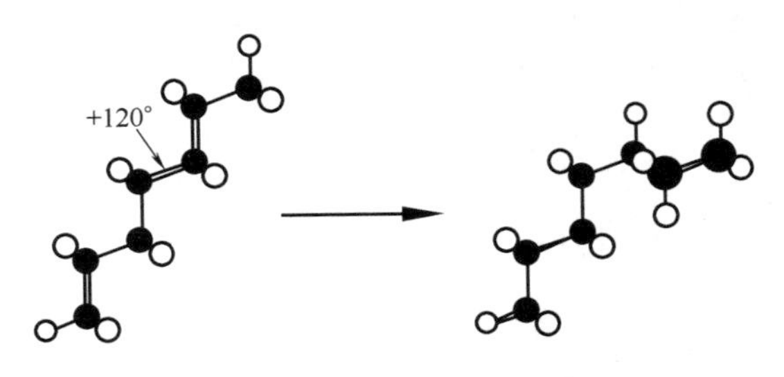

Figure 4.7 Examples of conformational states of a few repeating units of a polyethylene chain. The right-hand form is generated by 120° torsion about the single bond indicated by the arrow

A conformational state refers to the stereostructure of a molecule defined by its sequence of bonds and torsion angles. The change in shape of a given molecule due to torsion about single (sigma) bonds is referred to as a change of conformation (Figure 4.7)① . Double and triple bonds, which in addition to the rotationally symmetric sigma bond also consist of one or two rotationally asymmetric pi bonds, permit no torsion. There are only small energy barriers, from a few to 10 kJ mol^{-1}, involved in these torsions. The multitude of conformations in polymers is very important for the behaviour of polymers. The rapid change in conformation is responsible for the sudden extension of a rubber polymer on loading and the extraordinarily high ultimate extensibility of the network②. The high segmental flexibility of the molecules at high temperatures and the low flexibility at low temperatures is a very useful signature of a polymer.

1. New word

asymmetric[ˌæsiˈmetrɪk] *adj.* 不对称的，非对称的

torsion[ˈtɔːʃ(ə)n] *n.* 扭转，扭曲，转矩，[力]扭力

segmental[segˈmentəl] *adj.* 部分的

polybutadiene[pɔliˌbjuːtəˈdaɪiːn] *n.* 聚丁二烯

2. Phrases and expressions

configurational state　构型状态

atactic polymer　无规聚合物，不规则排列聚合物

isotactic polymer　等规聚合物，［高分子］全同立构聚合物，全同聚合物

3. Notes to the text

① The change in shape of a given molecule due to torsion about single (sigma) bonds is referred to as a change of conformation.　由于单个键(sigma)的扭转而引起的分子形状的变化称为构象的变化。

② The rapid change in conformation is responsible for the sudden extension of a rubber polymer on loading and the extraordinarily high ultimate extensibility of the network.　构象的快速变化导致了橡胶聚合物在载荷作用下的突然扩展和橡胶分子链网络的超高终极可扩展性。

Lesson 5 Structure of Synthesis

As an appreciation for the molecular structure of polymers was gained, three major structural categories emerged. These are illustrated schematically in Figure 5. 1.

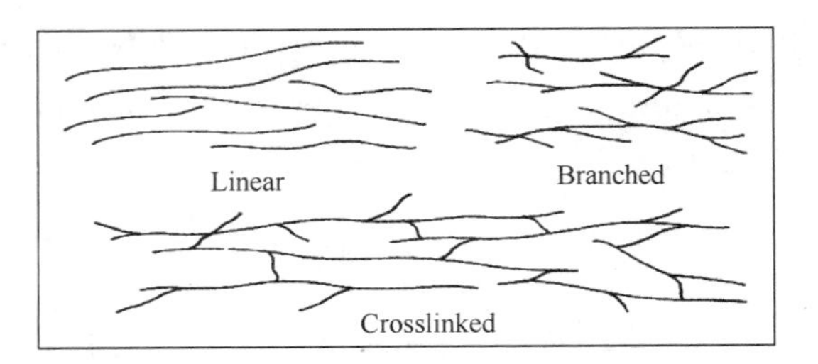

Figure 5. 1 Schematic diagram of polymer structures

A. Linear

If a polymer is built from strictly difunctional monomers, the result is a *linear* polymer chain. A scale model of a typical linear polymer molecule made from 0. 5-cm-diameter clothesline would be about three meters long. This isn't a bad analogy: The chains are long, flexible, essentially one-dimensional structures. The term linear can be somewhat misleading, however, because the molecules don't necessarily assume a geometrically linear conformation①.

B. Branched

If a few points of tri(or higher) functionality are introduced(either intentionally or through side reactions) at random points along linear chains, *branched* molecules result②. Branching can have a tremendous influence on the properties of polymers through steric(geometric) effects.

Under specialized conditions, branches of repeating unit B may be "grafted" to a backbone of linear A. This structure is known as a *graft copolymer*:

```
           B                       B
         B                       B
       B                       B
     B                       B
AAAA------------A------------A
                 B
                   B
                     B
```

The graft copolymer above would be called poly(A-*g*-B), the backbone repeating unit being the one listed first.

C. Crosslinked or Network

As the length and frequency of the branches on polymer chains increases, the probability that the branches will finally reach from chain to chain, connecting them together, becomes greater③. When all the chains are finally connected together in three dimensions by these crosslinks, the entire polymer mass becomes one tremendous molecule, a *crosslinked* or *network* polymer. In a truly crosslinked polymer mass, all the atoms are connected to one another by covalent bonds. The polymer in a bowling ball, for example, has a molar mass on the order of 10^{27} g/mol. Remember this the next time someone suggests that individual molecules are too small to be seen with the naked eye④.

1. New words

appreciation[əˌpriːʃiˈeiʃən] *n.* 理解，领会
schematically[skiˈmætikli] *adv.* 示意地，大略地
linear[ˈliniə] *adj.* 线型的
diameter[daiˈæmitə] *n.* 直径
clothesline[ˈkləuðzlain] *n.* 晾衣绳
flexible[ˈfleksəbl] *adj.* 柔韧的
essentially[iˈsenʃəli] *adv.* 本质上
somewhat[ˈsʌmwɔt] *adv.* 稍微，有点
misleading[misˈliːdiŋ] *adj.* 使人误解的
geometrically[ˌdʒiəˈmetrikəli] *adv.* 几何学上
branched[bræntʃd] *adj.* 支化的
intentionally[inˈtenʃənli] *adv.* 有意地，故意地
tremendous[triˈmendəs] *adj.* 巨大的，惊人的
property[ˈprɔpəti] *n.* 性能，性质
backbone[ˈbækbəun] *n.* 主链
crosslinked[ˈkrɔsliŋkd] *adj.* 交联的
network[ˈnetwəːk] *n.* 网络
frequency[ˈfriːkwənsi] *n.* 频率
branch[bræntʃ] *n.* 分枝（支链）
crosslink[ˈkrɔsliŋk] *vt.* 交联；*n.* 交联点

2. Phrases and expressions

schematic diagram 原理图，示意图
one-dimensional structure 一维结构
linear conformation 线形构象
side reaction 副反应
steric (geometric) effect 立体（几何）效应
graft copolymer 接枝共聚物
repeating unit 重复单元
three dimensions 三维空间
bowling ball 保龄球
on the order of 大约，数量级为
naked eye 肉眼

3. Notes to the text

① The term linear can be somewhat misleading, however, because the molecules don't necessarily assume a geometrically linear conformation. 然而“线型的”这一术语多少有些使人误解，因为分子未必是几何学上的线型形态。

② If a few points of tri (or higher) functionality are introduced (either intentionally or through side reactions) at random points along linear chains, branched molecules result. 如果沿线型分子链，在随机的若干个点上引入（无论是有意引入还是通过副反应引入）一些三（或更高）官能度（单体），就形成了支化分子。

③ As the length and frequency of the branches on polymer chains increases, the probability that the branches will finally reach from chain to chain, connecting them together, becomes greater. 当聚合物分子链上支链的长度和密度增加时，支链从一条链延伸到另外一条链，并将它们最终连接到一起的几率就变大了。

④ The polymer in a bowling ball, for example, has a molar mass on the order of 10^{27} g/mol. Remember this the next time someone suggests that individual molecules are too small to be seen with the naked eye. 当下次有人提出单个分子很小，无法用肉眼观察到时，不要忘记这个事实，例如用来制作保龄球的聚合物的摩尔质量大约是 10^{27} g/mol。

4. Exercises

(1) Try to discuss the formation conditions of linear, branched, crosslinked and network polymers according to the text.

(2) HIPS (High Impact Polystyrene), which is made of butadiene and styrene, is a typical graft copolymer. Assuming that the backbone of HIPS is absolutely comprised of butadiene, try to draw up the schematic diagram of macromolecular structure of HIPS.

(3) Given supplies of acrylic acid (Ⅰ), adipic acid (Ⅱ), and propylene glycol (Ⅲ)

$H_2C{=\!=}CHCOOH$ (Ⅰ)　　$HOOC\!\left(CH_2\right)_4COOH$ (Ⅱ)　　$HO\!\left(CH_2\right)_3OH$ (Ⅲ)

a. Show the repeating unit(s) of all the linear homopolymers that could be produced.

b. Describe three different procedures by which crosslinked polymers could be produced (again using only Ⅰ, Ⅱ, and Ⅲ).

Reading Material

Bonding in Polymers

TYPES OF BONDS

Various types of bonds hold together the atoms in polymeric materials, unlike in metals, for example, where only one type of bond exists. These types are (1) primary covalent, (2) hydrogen bond, (3) dipole interaction, (4) van der Waals and, (5) ionic. Hydrogen bonds, dipole interactions, van der Waals bonds, and ionic bonds are known collectively as secondary bonds. The distinctions are not always clear-cut; that is, hydrogen bonds may be considered as the extreme of dipole interactions①. Bond parameters of different types of bonds are shown in Table 5. 1.

Table 5. 1　Bond parameters

Bond type	Interatomic distance r_m/nm	Dissociation energy ε/(kcal/mol)
Primary covalent	0. 1 ~ 0. 2	50 ~ 200
Hydrogen bond	0. 2 ~ 0. 3	3 ~ 7
Dipole interaction	0. 2 ~ 0. 3	1. 5 ~ 3
van der Waals bond	0. 3 ~ 0. 5	0. 5 ~ 2
Ionic bond	0. 2 ~ 0. 3	10 ~ 20

BONDING AND RESPONSE TO TEMPERATURE

In linear and branched polymers, only the secondary bonds hold the individual polymer chains together (neglecting temporary mechanical entanglements). Thus, as the temperature is raised, a point will be reached where the forces holding the chains together become insignificant, and the chains are then free to slide past one another, that is, to *flow* upon the application of stress②. Therefore, *linear* and *branched* polymers are generally *thermoplastic*. The crosslinks in a crosslinked polymer, on the other hand, are held together by the same primary covalent bonds as are the main chains. When the thermal energy exceeds the dissociation energy of the primary covalent bonds, both main-chain and crosslink bonds fail randomly, and the polymer degrades. Hence, *crosslinked polymers are thermosetting*.

ACTION OF SOLVENTS

The action of solvents on polymers is in many ways similar to that of heat. Appropriate solvents, that is, those that can form strong secondary bonds with the polymer chains, can penetrate, replace the interchain secondary bonds, and thereby pull apart and dissolve linear and branched polymers③. The polymer-solvent secondary bonds cannot overcome primary valence crosslinks, however, so crosslinked polymers are not soluble, although they may swell considerably. (Try soaking a rubber band in toluene overnight.) The amount of swelling is, in fact, a convenient measure of the extent of crosslinking. A lightly crosslinked polymer such as a rubber band will swell tremendously, while one with extensive crosslinking, for example, an ebonite ("hard rubber") bowling ball, will not swell noticeably at all.

1. New words

metal['metl] *n.* 金属;*adj.* 金属制的

interaction[ˌintər'ækʃən] *n.* 相互作用,相互影响

ionic[ai'ɔnik] *adj.* 离子的

extreme[iks'tri:m] *n.* 极端,极端的;*adj.* 末端的

neglect[ni'glekt] *vt.* 疏忽,忽视;*n.* 疏忽,忽略

temporary['tempərəri] *adj.* 暂时的,临时的

insignificant[ˌinsig'nifikənt] *adj.* 无关紧要的,可忽略的

slide[slaid] *vi.* 滑,跌落

exceed[ik'si:d] *vt.* 超过,胜过

parameter[pə'ræmitə] *n.* 参数,参量

solvent['sɔlvənt] *n.* 溶剂

penetrate['penitreit] *vt.* ;*vi.* 穿透,渗透

interchain[ˌintə'tʃein] *adj.* 链间的

swell[swel] *vi.* 溶胀

soak[səuk] *vi.* 浸润

toluene['tɔljui:n] *n.* 甲苯

ebonite['ebənait] *n.* 硬质橡胶

noticeably['nəutisəbli] *adv.* 显著地,明显地

perduren[pə:'dju:rən] *n.* 硫化橡胶

2. Phrases and expressions

hydrogen bond 氢键
van der Waals 范德瓦尔斯(范德华)
mechanical entanglement 机械缠结
primary covalent bond 主价键
thermal energy 热能
dissociation energy 离解能
appropriate solvent 适当溶剂(良溶剂)
extent of crosslinking 交联度

3. Notes to the text

① The distinctions are not always clear-cut; that is, hydrogen bonds may be considered as the extreme of dipole interactions. 差别并不总是十分明确的;也就是说,氢键可以看成是偶极作用的极端情况。

② Thus, as the temperature is raised, a point will be reached where the forces holding the chains together become insignificant, and the chains are then free to slide past one another, that is, to flow upon the application of stress. 因此,当温度升高到某一点时,将分子链束缚在一起的作用力变得微乎其微,分子链之间可以自由滑动,也就是说,可以在外力的作用下发生流动。

③ Appropriate solvents, that is, those that can form strong secondary bonds with the polymer chains, can penetrate, replace the interchain secondary bonds, and thereby pull apart and dissolve linear and branched polymers. 适当的溶剂,即能与聚合物分子链形成强次价键的溶剂,能够渗透到分子链间,取代分子链间的次价键,从而拆散并溶解线型和支化聚合物。

Lesson 6 Chain Conformation

A polymer molecule can take many different shapes (conformations) primarily due its degree of freedom for rotation about σ bonds. Studies of the heat capacity of ethane (CH_3—CH_3) indicate that the bond linking the carbon atoms is neither completely rigid nor completely free to rotate. Figure 6.1 shows the different rotational positions of ethane as viewed along the C—C bond. The hydrogen atoms repel each other, causing energy maxima in the eclipsed position and energy minima in the stable staggered position. The torsion angle may be defined as in Figure 6.1, $\Phi = 0$ for the eclipsed position and $\Phi > 0$ for clockwise rotation round the further carbon atom. Some authors set $\Phi = 0$ for the staggered position. In both cases, the value of φ is independent of the viewing direction (turning the whole molecule round).

Figure 6.2 shows the conformational energy plotted as a function of the torsion angle and the energy difference between the stable staggered positions. The energy barrier (eclipsed) is equal to 11.8 kJ/mol which may be compared with the thermal energy at room temperature, $RT \approx 8.31 \times 300 J/mol \approx 2.5 kJ/mol$.

The alkane with additional two carbon atoms, n-butane (CH_3—CH_2—CH_2—CH_3), has different stable conformational states, referred to as trans (T) and gauche (G and G′), as shown in Figure 6.3. The conformational energy 'map' of n-butane is shown in Figure 6.4. The energy difference between the trans and gauche states is $2.1 \pm 0.4 kJ/mol$. Calculations and experiments have shown that there is an angular displacement by $5 \sim 10°$ of the gauche states from their 120° angle towards the trans state, i.e. the gauche states are located at $110 \sim 115°$ from the trans state. The energy barrier between the trans and the gauche states is 15kJ/mol. The energy barrier between the two gauche states is believed to be very high, but its actual value is not precisely known.

Normal pentane has two rotational bonds and hence potentially nine combinations, but only six of them are distinguishable: TT, TG, TG′, GG, G′G′ and GG′. The conformations GT, G′T and G′G are identical with TG, TG′ and GG′. Two pairs of mirror-images are present, namely TG and TG′ and GG and G′G′. The energy for the conformation GG′ is much greater than predicted from the data presented in Figure 6.4 because of strong steric repulsion of the two CH_3 groups separated by three CH_2 groups (Figure 6.5). The dependence of the potential energy of one σ bond on the actual torsion angle of the nearby bonds is referred to as a second-order interaction.

The rotational isomeric state approximation, which is a convenient procedure for

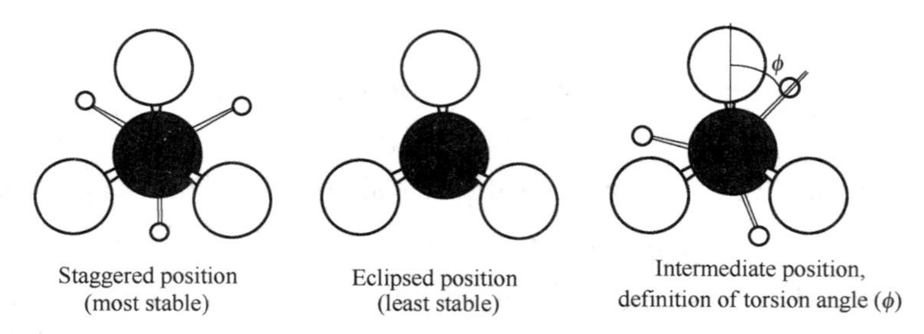

Figure 6. 1 Rotational isomers of ethane from a view along the c—c bond: carbon-shaded; hydrogen-white

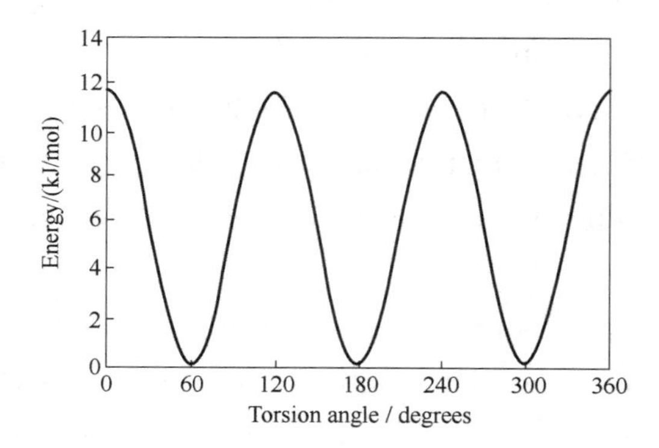

Figure 6. 2 Conformational energy of ethane as a function of torsion angle

dealing with the conformational states of polymers, was introduced by Flory[①]. Each molecule is treated as existing only in discrete torsional angle states corresponding to the potential energy minima, i. e. to different combinations of T, G and G′. Fluctuations about the minima are ignored. This approximation means that the continuous distribution over the torsional angle space Φ is replaced by a distribution over many discrete states. This approximation is well established for those bonds with barriers substantially greater than the thermal energy (RT).

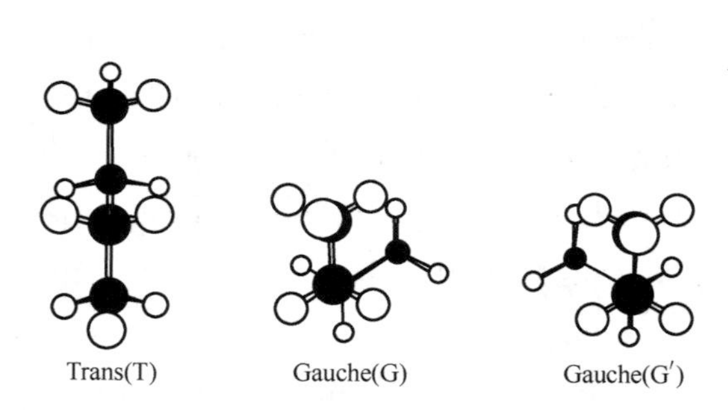

Figure 6. 3 Conformational states of n-butane. Note that the views of the gauche conformers are along the middle carbon carbon bond. Carbon-shaded; hydrogen-white

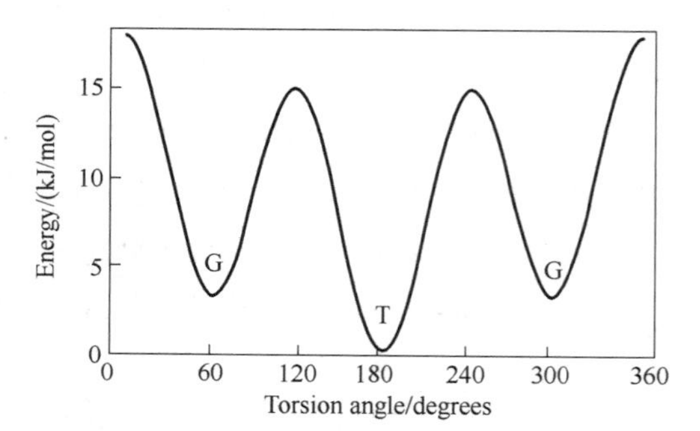

Figure 6. 4 Conformational energy of n-butane as a function of torsion angle of the central carbon-carbon bond. The outer carbon-carbon bonds are assumed to be in their minimum energy states (staggered positions)

Let us now consider an alkane with n carbons. The question is how many different conformations this molecule can take. The molecule with n carbons has $n-1$ σ main-chain bonds. The two end bonds do not contribute to different conformations, whereas each of the other carbon-carbon bonds is in one of the three rotational states T, G and

G′. The number of different conformations following this simple scheme is $3n$-3. A typical polymer molecule may have 10000 carbons and thus 39997 ≈ 104770 conformations, i. e. an enormously large number of states. However, this treatment is oversimplified. First, due to symmetry, the number of distinguishable conformations is less than 39997, although it is correct in order of magnitude. Second, the energy of certain conformations is very high, e. g. those containing GG′, giving them a very low statistical weight. The energy map shown in Figure 6.4 is of limited applicability for predicting the probability of conformations in polyethylene. The interdependence of the torsion angle potentials, as demonstrated in the high-energy GG′ sequence, has to be considered. These higher-order interactions are discussed later in this chapter.

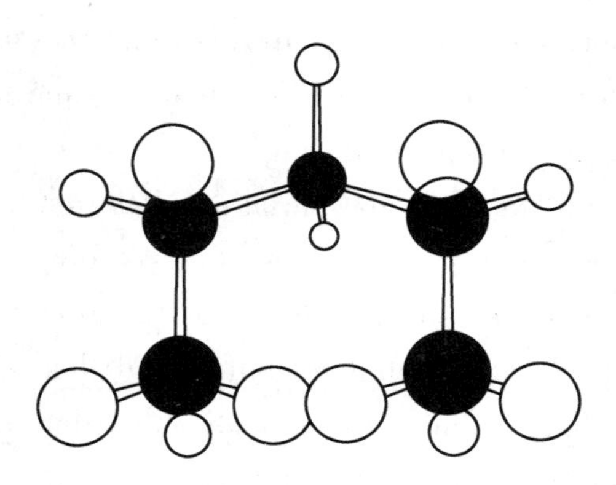

Figure 6.5 Illustration of the steric repulsion in the high-energy GG′ conformer in n-pentane: carbon-shaded; hydrogen-white

Polymer chains exhibit in several cases a random chain conformation, i. e. a random distribution of trans and gauche states. Figure 6.6 illustrates the idea of a random (so-called Gaussian) chain. Random macromolecular chains are found in solutions of polymers in good solvents, in polymer melts and probably also in glassy amorphous polymers. Crystalline polymers, on the other hand, consist of long sequences of bonds with a regular arrangement of energetically preferred chain conformations interrupted by chain folds or by statistical sequences.

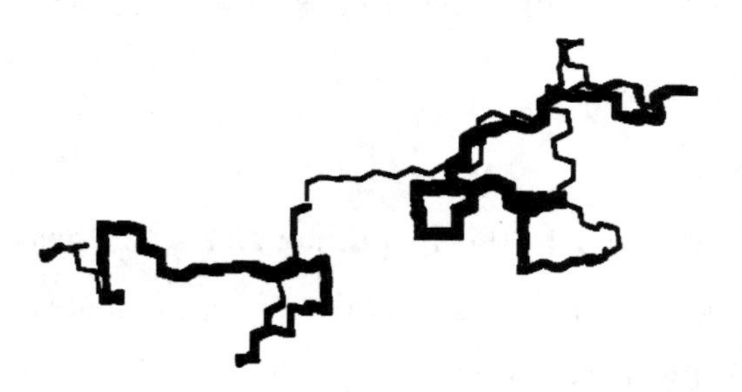

Figure 6.6 Gaussian chain

The first part of this chapter deals with the statistics of the Gaussian chain. Expressions for the characteristic dimension of the random chain (average end-to-end distance or radius of gyration) are derived as a function of molar mass, chain flexibility and temperature.

The particularly simple relationships between the average end-to-end distance of the random coil and the chain length that are derived in the previous are valid under the ideal solution conditions referred to as theta conditions. The dimension of the unperturbed polymer chain is only determined by the short-range effects and the chain behaves as a 'phantom' chain that can intersect or cross itself freely. It is important to note that these conditions are also met in the pure polymer melt, as was first suggested by Flory (the so-called Flory theorem) and as was later experimentally confirmed by small-angle neutron scattering.

The statistical variation of the end-to-end distance is considered in so-called random flight analysis. This analysis forms the basis for one of the most prominent theories in polymer physics, the theory of rubber elasticity.

It is essential to understand that an ensemble of random chains can only be described by means of a spatial distribution function②. Two different measures of the random chain are commonly used: (a) the end-to-end distance (r); and (b) the radius of gyration (S). The former is simply the distance between the chain ends. The radius of gyration is defined as the root-mean-square distance of the collection of atoms from their common centre of gravity:

$$S^2 = \frac{\sum_{i=1}^{n} m_i r_i^{-2}}{\sum_{i=1}^{n} m_i} \tag{6.1}$$

where r_i is the vector from the centre of gravity to atom i. Debye showed many years ago that for large values of n, i. e. for polymers, the following relationship holds between the second moment of the mean values:

$$\langle S^2 \rangle = \frac{\langle r^2 \rangle}{6} \tag{6.2}$$

Random coils can thus be characterized by either of the two dimensions, the average end-to-end distance or the radius of gyration. The average end-to-end distance is used in the rest of this chapter to characterize the random chain.

1. New word

staggered position ['stægəd] *adj.* 错列开的位置

gauche (conformation) 邻位交叉(构象), 扭曲(构象)

theorem ['θiərəm] *n.* [数]定理

gyration [dʒai'reʃən] *n.* 旋转, [力]回转

2. Phrases and expressions

eclipsed position 重叠的位置

torsional angle 扭转角

Gaussian chain 高斯链

3. Notes to the text

① The rotational isomeric state approximation, which is a convenient procedure for dealing with the conformational states of polymers, was introduced by Flory.

弗罗里提出了旋转异构态近似理论,这是一种处理聚合物构象态的简便方法。

② It is essential to understand that an ensemble of random chains can only be described by means of a spatial distribution function.

至关重要是必须理解，只能借助空间分布函数来描述随机链集合。

4. Exercise

Demonstrate rotational isomers of ethane from a view along the c-c bond according to the text.

Reading Material

Chains with Preferred Conformation

Polymer molecules are found in a preferred conformational state in crystals. The experimental techniques for determining the preferred conformation are mainly X-ray and electron diffraction. The difficult determination of the crystal unit cell must be followed by further molecular mechanical modelling to establish the exact chain conformation①.

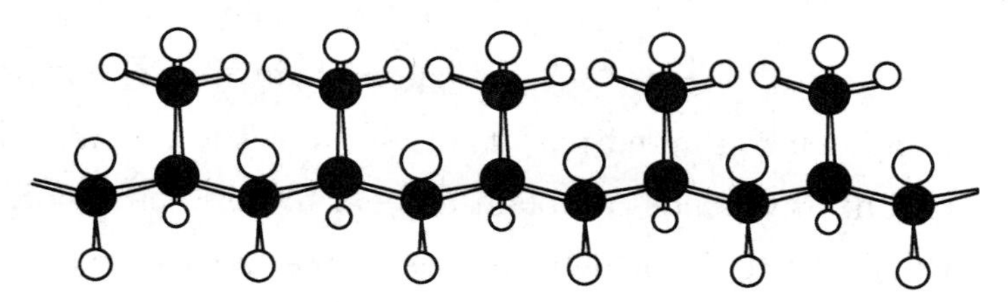

Figure 6.7 Isotactic polypropylene in all-trans conformation showing the steric problem associated with the pendant methyl groups.

Polyethylene obviously shows the simplest polymer structure. The all-trans conformation is energetically the most stable conformation and has been established by numerous diffraction experiments②. For polymers with pendant side groups, e. g. isotactic polypropylene (iPP) and isotactic polystyrene (iPS), the extended all-trans conformation is of high energy due to steric repulsion of the side groups (Figure 6.7).

For iPP, two sequences, /TG/TG/TG/TG/ and /G′T/G′T/G′T/G′T/, have the same minimum conformational energy. Both conformations produce helices. Three polymer repeating units produce a repeating unit in one turn of the helix. This kind of helix is denoted 3_1. The two conformations have different pitches. A view along the helical axis is shown in Figure 6.8. Other isotactic polymers, e. g. iPS, also prefer the 3_1 TG helix for the same reason as does iPP.

A polymer molecule can adopt many different shapes primarily due to its degree of freedom for torsion about σ bonds. These states are referred to as conformations. A polymer molecule in a solution, in the molten state and probably also in the glassy, amor-

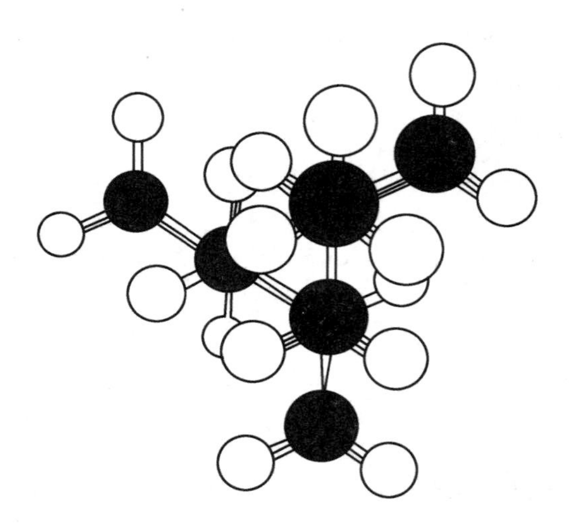

Figure 6.8 View along helical axis of 3_1, helix of isotactic polypropylene. The cross-section of the backbone part of the molecule is triangular and the pendant methyl groups are directed out from the corners of the triangle.

phous state, can be characterized as a random coil.

It has been experimentally shown that the second moment of the end-to-end distance ($< r^2 >_0$) of unperturbed polymer chains, which only appear under so-called theta conditions, is proportional to the number of bonds (n) and the square of the length of each bond(l):

$$(< r^2 >)_0 = Cnl^2 \qquad (6.3)$$

Where C is a constant, which depends on the segmental flexibility of the polymer.

This kind of relationship can be derived on the basis of very simple models. For polyethylene the values for C are for a freely jointed chain $C = 1$, for a freely rotation chain $C = 2$, and for a chain with hindered rotation $C > 2$. Flory showed that it was sufficient to consider the energetics of the torsion about two nearby bonds to obtain agreement between predicted and experimental values($C = 6.8$ at 410 K for polyethylene in the theta state).

The unperturbed state, ie. the state of a polymer under theta conditions, is characterized by the absence of long-range interactions. The segments of a molecule under theta conditions are arranged in a way which indicates that they do not'sense' the other segments of the same molecule. The molecules behave like'ghosts' or'phantoms' and are sometimes also referred to as phantom chains. Flory proposed that the spatial extension of polymer molecules in the molten state is the same as in the theta solvent and that the same simple equation [eq. (6.3)] between $< r^2 >_0$ and chain length (n) should hold. Small-angle neutron scattering data were available may years later and supported the Flory theorem. In good solvents where, in addition to short-ranged interactions, long-ranged interactions also play a role, the perturbed state can be described by the following equation:

$$< r^2 > \propto n^{6/5} \qquad (6.4)$$

The phantom(unperturbed) polymer chain be can represented by a hypothetical chain with $n' = r/C$ feely jointed segments each of length $l' = Cl$. If n and l are replaced by n' and l' in the equation for the feely jointed chains, eq. (6.3) is obtained.

For phantom chains the distribution function of the end-to-end distance is Gaussian, taking the form:

$$P(x,y,z)\,dx\,dy\,dz = \left(\frac{3}{2\pi < r^2 >_0}\right)^{3/2} \times \exp(-3r^2/2 < r^2 >_0)\,dx\,dy\,dz \qquad (6.5)$$

This Gaussian expression is fundamental to the expression statistical mechanical theory of rubber elasticity.

Polymer chains in crystals take their preferred conformation, i. e. their low-energy state. Linear polymers with small pendant groups, e. g. polyethylene, exhibit an extended all-trans conformation. Isotactic polymers with the repeating unit—CH_2—CHX—exhibit a helical structure if the X group is sufficiently large (i. e. If X is a methyl group or larger). Even linear polymers with no large pendant group may, due to electrostatic repulsion between nearby dipoles, form a helical structure. Polyoxymethylene belongs to this category.

1. New word

atactic[e'tæktɪk] *adj.* 不规则的,[有化]无规立构的

isotactic[ˌaisəu'tæktik] *adj.* 全同立构的,等规立构的

polyoxymethylene[ˌpɔliˌɔksi'meθili:n] *n.* 聚甲醛

helix['hilɪks] *n.* 螺旋

2. Phrases and expressions

poly(vinyl alcohol) 聚(乙烯醇)

pendant group 侧基

steric repulsion 位阻排斥力

3. Notes to the text

① The difficult determination of the crystal unit cell must be followed by further molecular mechanical modelling to establish the exact chain conformation. 在对晶体单元进行困难的测定之后,必须用进一步的分子力学模型来建立精确的链构象。

② The all-trans conformation is energetically the most stable conformation and has been established by numerous diffraction experiments.

全反式构象在能量上是最稳定的构象,是经过大量地衍射实验建立起来的。

Lesson 7 Polymer Morphology

We know that the geometric arrangement of the atoms in polymer chains can exert a significant influence on the properties of the bulk polymer. To appreciate why this is so, the subject of polymer morphology, the structural arrangement of the chains in the polymer, is introduced here.

Although the precise nature of crystallinity in polymers is still under investigation, a number of facts have long been known about the requirements for polymer crystallinity[①]. First, an ordered, regular chain structure is necessary to allow the chains to pack into an ordered, regular, three-dimensional crystal lattice. Thus, stereo regular polymers are more likely to be crystalline than those that have irregular chain structures. Irregularly spaced, protruding side groups hinder crystallinity. Second, no matter how regular the chains, the secondary forces holding the chains together in the crystal lattice must be strong enough to overcome the disordering effect of thermal energy; so hydrogen bonding or strong dipole interactions promote crystallinity, and other things being equal, raise the crystalline melting point.

X-ray studies show that there are numerous polymers that do not meet the above criteria and show no traces of crystallinity; that is, they are completely amorphous. In contrast to the regular, ordered arrangement in a crystal lattice, the chains in an amorphous polymer mass assume a more-or-less random, twisted, entangled, "balled-up" configuration[②]. A common analogy is a bowl of cooked spaghetti. A better analogy, in view of the constant thermal motion of the chain segments, is a bowl of wriggling snakes. The latter analogy forms the basis of a currently popular quantitative model of polymer behavior known as reputation (for reptiles) theory.

The first direct observations of the nature of polymer crystallinity resulted from the growth of single crystals from dilute solution. Either by cooling or evaporation of solvent, thin, pyramidal or plate-like polymer crystals (lamellae) was precipitated from dilute solutions (Figure 7.1). These crystals were on the order of 10 μm along a side and only about 0.01 μm thick. This was fine, except that X-ray measurements showed that the polymer chains were aligned perpendicular to the large flat faces of the crystals, and it was known that the extended length of the individual chains was on the order of 0.1 μm. How could a chain fit into a crystal one-tenth its length? The only answer is that the chain must fold back on itself, as shown in Figure 7.1. Two competing models of chain folding are illustrated.

This folded-chain model has been well substantiated for single polymer crystals. The lamellae are about 50 to 60 carbon atoms thick, with about five carbon atoms

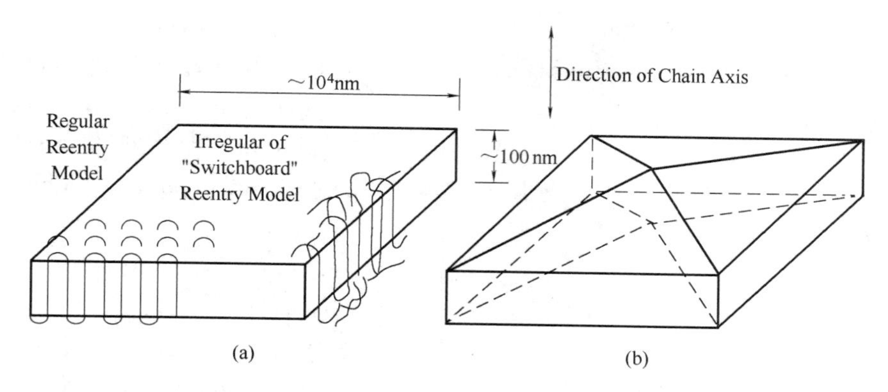

Figure 7.1 Polymer single crystals (a) flat lamellae; (b) pyramidal lamellae. Two concepts of chain reentry are illustrated

in a direct reentry fold. The atoms in a fold, whether direct or indirect reentry, can never be part of a crystal lattice.

It is now well established that similar lamellar crystallites exist in bulk polymer samples crystallized from the melt, although the lamellae may be up to 1 μm thick. Recent results support the presence of a third, interfacial region between the crystalline lamellae and the amorphous phase. This interfacial phase can make up some 10 to 20% of the material. Furthermore, there doesn't seem to be much, if any, direct-reentry folding of chains in bulk-crystallized lamellae. This is illustrated in Figure 7.2. Orientation of the lamellae along with additional orientation and crystallization in the interlamellar amorphous regions, as in the fringed-micelle model, is usually invoked to explain the increase in the degree of crystallinity with drawing.

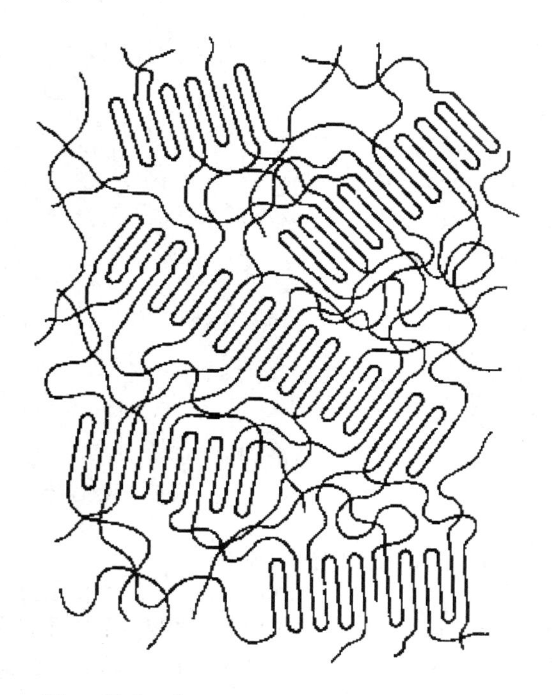

Figure 7.2 Compromise model showing folded-chain lamellae tied together by interlamellar amorphous chains

1. New words

morphology [mɔː'fɔlədʒi] *n.* 形态，形态学

geometric [dʒi'ɔmətri] *adj.* 几何的

exert [ig'zəːt] *vt.* 发挥，施加

bulk [bʌlk] *n.* 大小，体积，大批；*vt.* 显得大，显得重要；*adj.* 大批合计的

crystallinity ['kristəlainiti] *n.* 结晶

crystal ['kristəl] *n.* 水晶，结晶（体）；*adj.*

水晶制的,水晶般的,透明的
three-dimensional[θri:di'menʃənəl] *adj.* 三维的,立体的
lattice['lætis] *n.* 格子
protruding[prə:'tru:diŋ] *adj.* 凸,突出
hydrogen['haidrədʒən] *n.* 氢,氢气
bond[bɔnd] *vi.* 粘接
criterion[krai'tiəriən] *n.* 标准(*Pl.* -criteria)
amorphous[ə'mɔ:fəs] *adj.* 无定型的,非晶的
twist[twist] *vi.* 蜷曲
entangle[in'tæŋgl] *vt.* 使纠缠,卷入,使混乱
spaghetti[spə'geti] *n.* 意大利面条
motion['məuʃən] *n.* 动力,运动
wriggling['rigliŋ] *adj.* 蠕动
dilute[dai'lju:t] *adj.* 稀,经稀释的
evaporation[iˌvæpə'reiʃən] *n.* 干燥,蒸发
pyramidal[pi'ræmidl] *adj.* 锥体的
precipitate[pri'sipiteit] *vi.* 沉淀,析出
align[ə'lain] *vi.* 排列,排成一行
perpendicular[ˌpə:pən'dikjulə] *adj.* 直立的,垂直
substantiate[səb'stænʃieit] *vt.* 证明,证实
reentry[ri:'entri] *n.* 折返
crystallize['kristəlaiz] *vt.* 使……结晶
interfacial[ˌintə(:)'feiʃəl] *adj.* 界面的
crystalline['kristəlain] *adj.* 水晶制的,透明的
orientation[ˌɔ:rien'teiʃən] *n.* 取向
crystallization['kristəlai'zeiʃən] *n.* 结晶化
interlamellar[ˌintə(:)lə'melə] *adj.* 层间的
invoke[in'vəuk] *vt.* 援引
drawing['dʒɔ:iŋ] *n.* 拉伸

2. Phrases and expressions

three-dimensional crystal lattice　三维晶格
crystal lattice　晶格,晶体点阵
"balled-up" configuration　球状构型
amorphous phase　非晶相
interfacial phase　界面相
fringed-micelle model　缨状微束模型

3. Notes to the text

① Although the precise nature of crystallinity in polymers is still under investigation, a number of facts have long been known about the requirements for polymer crystallinity.　虽然对聚合物结晶的确切性质仍然在进行研究,但是人们早就知道了关于聚合物结晶必要条件的大量事实。

② In contrast to the regular, ordered arrangement in a crystal lattice, the chains in an amorphous polymer mass assume a more-or-less random, twisted, entangled, "balled-up" configuration.　与晶格中规则的和有秩序的排列相反,非晶聚合物的链在一定程度上是无规则的、扭曲的、缠绕的和球状构型。

4. Exercises

(1)________ are not hinder crystallinity?

a. Irregularly spaced groups　b. Protruding side groups

c. Regular groups

(2) Suppose that you know the theoretical enthalpy of melting for a 100% crystalline polymer. Show how you could obtain the degree of crystallinity for an actual sample by measuring its enthalpy of melting.

(3) What are the two analogies (a bowl of cooked spaghetti and a bowl of wriggling snakes) illustrated for?

(4) Try to explain the increase in the degree of crystallinity with drawing.

Reading Material

Amorphous Polymers

Let us consider a polymer which is in its molten state. What may happen when it cools down? The two possibilities are shown in Figure 7.3. The polymer may either crystallize (route a) or cool down to its glassy, amorphous state (route b). The temperature at which the slope in the specific volume-temperature graph (route b) changes is referred to as the glass transition temperature, Tg. Rigid-rod polymers, i.e. polymers with very inflexible groups in the backbone chain or in side chains, may form liquid-crystalline states.

Cooling of a liquid following routes a crystallization or forming a glassy amorphous structure

What molecular factors determine whether a polymer will crystallize or not? The regularity of the polymer is the key factor: isotactic polypropylene crystallizes, whereas atactic polypropylene does not. Atactic polymers generally do not crystallize with two exceptions:

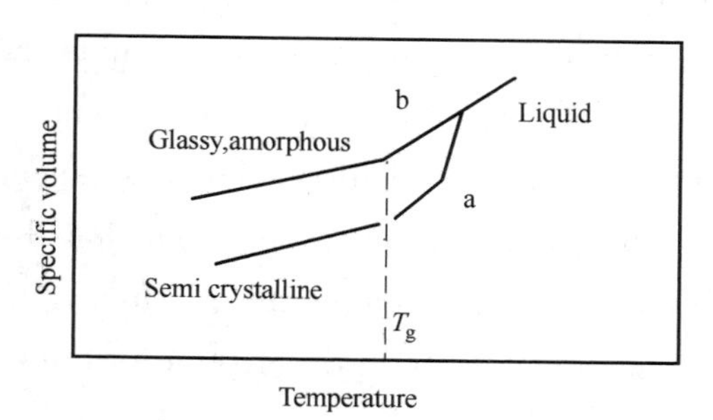

Figure 7.3 cooling of a liquid following routes a (crystallization) or b (forming a glassy amorphous structure)

(1) The X group in ($—CH_2—CHX—$), is very small allowing regular packing of the chains regardless of whether the different pendant groups are randomly placed. Poly (vinyl alcohol) with its small hydroxyl X groups is a good example of this kind of polymer.

(2) The X group forms a long regular side chain. Side-chain crystallization may occur provided that the pendant groups are of sufficient length.

Random copolymers are incapable of crystallizing except when one of the constituents is at a significantly higher concentration than the other constituent①. (Linear low-density polyethylene with a crystallinity of about 50% contains 98.5 mol % of methylene units and 1.5 mol % of CHX units, where X is $—CH_2CH_3$ or a longer homologue. Polymers which potentially crystallizable may be quenched to a glassy amorphous state. Polymers with large side groups having an inflexible backbone chain, are more readily quenched to a glassy, fully amorphous polymer than flexible polymers such as polyethylene. Figure7.4 shows schematically the effects of polymer structure and cooling rate on the solidified polymer structure.

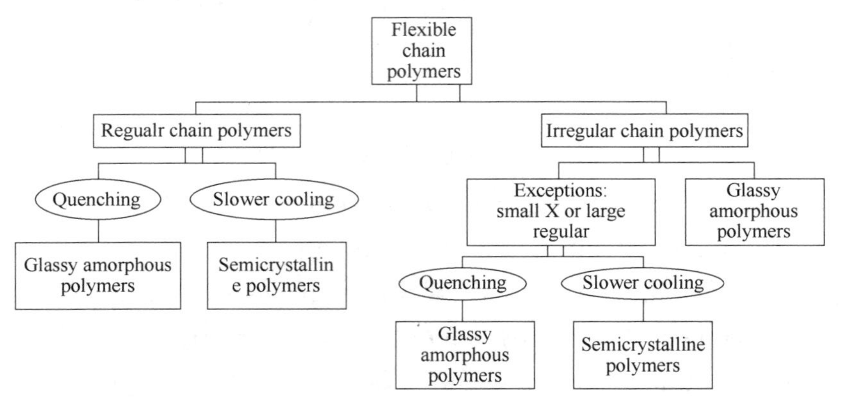

Figure 7.4 The effect of molecular and thermal factors on the structure of the solidified polymer

Cooling past the glass transition temperature is accompanied by a dramatic change in the mechanical properties. The elastic modulus increases by a factor of 1000 when the polymeric liquid is cooled below T_g and the modulus of the glassy polymers is relatively insensitive to changes in molar mass and repeating unit structure[②]. The actual value of T_g is, however, very dependent on the repeating unit, the molecular architecture and the presence of low molar mass species.

It is appropriate to point out that the T_g value recorded in any given experiment is dependent on the temperature scanning rate. A statement about a T_g value should always be accompanied by a description of the experiment. The remainder of this chapter presents a summary of the non-equilibrium nature of the glassy material, the current theories of the glass transition, the mobility of the molecules in their glassy state and finally the structure of glassy, amorphous polymers.

1. New words

amorphous[ə'mɔːfɔs] *adj.* 无定形的，无组织的，[物]非晶形的

crystalline['krist(ə)lain] *adj.* 透明的，结晶的，水晶制的；*n.* 结晶质

glassy['glɑːsi] *adj.* 像玻璃的；光亮透明的，呆滞的

equilibrium[ˌiːkwi'libriəm] *n.* 均衡，平静，保持平衡的能力

2. Phrases and expressions

random copolymer 无规共聚物

isotactic polypropylene 全同立构聚丙烯

atactic polypropylene 无规立构聚丙烯
glassy amorphous state 玻璃无定形状态
elastic modulus 弹性模数

3. Notes to the text

① Random copolymers are incapable of crystallizing except when one of the constituents is at a significantly higher concentration than the other constituent.

无规共聚物不能结晶,除非其中一个组分比其他组分的含量高得多。

② The elastic modulus increases by a factor of 1000 when the polymeric liquid is cooled below T_g and the modulus of the glassy polymers is relatively insensitive to changes in molar mass and repeating unit structure. 当聚合物液体冷却到 T_g 以下时,弹性模量增加了 1000 倍,玻璃聚合物的模量对于摩尔质量和重复单元结构的变化相对不敏感。

Lesson 8 Molar Mass

The enormous size of polymer molecules gives the unique properties. It shows the influence of molar mass on the melting point of polyethylene. Low molar mass substances(oligomers) show a strong increase in melting point with increasing molar mass, whereas a constant melting point is approached in the polymer molar mass range①. Other polymers show a similar behaviour.

Other properties such as fracture toughness and Young's modulus show a similar molar mass dependence, with constant values approached in the high molar mass region. Polymer properties are often obtained in the molar mass range from 10000 to 30000g/mol. Rheological properties such as melt viscosity show a progressive strong increase with increasing molar mass even at high molar masses. High molar mass polymers are therefore difficult to process but, on the other hand, they have very good mechanical properties in the final products.

There is currently no polymerization method available that yields a polymer with only one size of molecules. Variation in molar mass among the different molecules is characteristic of all synthetic polymers. They show a broad distribution in molar mass. The molar mass distribution ranges over three to four orders of magnitude in many cases. The full representation of the molar mass distribution is currently only achieved with size exclusion chromatography, naturally with a number of experimental limitations. Other methods yield different averages. The most commonly used averages are defined as follows. The number average is given by:

$$\overline{M}_n = \frac{\sum_i N_i M_i}{\sum_i N_i} = \sum_i n_i M_i \tag{8.1}$$

where N_i is the number of molecules of molar mass M_i and n_i is the numerical fraction of those molecules.

The mass or weight average is given by:

$$\overline{M}_w = \frac{\sum_i N_i M_i^2}{\sum_i N_i M_i} = \frac{\sum_i W_i M_i}{\sum_i W_i} = \sum_i W_i M_i \tag{8.2}$$

where W_i is the mass of the molecules of molar mass M_i and W_i; is the mass fraction of those molecules.

The Z average($\overline{M}_z$) is given by:

$$\overline{M}_z = \frac{\sum_i N_i M_i^3}{\sum_i N_i M_i^2} \tag{8.3}$$

while the viscosity average($\overline{M}_v$)

$$\overline{M}_v = \left(\frac{\sum_i N_i M_i^{1+i}}{\sum_i N_i M_i}\right)^{1/\alpha} \tag{8.4}$$

where α is a constant that takes values between 0.5 and 0.8 for different combinations of polymer and solvent. The viscosity average is obtained by viscometry. The intrinsic viscosity is given by:

$$[\eta] = \lim_{c\to 0}\left(\frac{\eta - \eta_0}{c\eta_0}\right) \tag{8.5}$$

where c is the concentration of polymer in the solution, η_0 is the viscosity of the pure solvent and η is the viscosity of the solution. The viscosities are obtained from the flow-through times(t and t_0) in the viscometer:

$$\frac{\eta}{\eta_0} \approx \frac{t}{t_0} \tag{8.6}$$

and the intrinsic viscosity is converted to the viscosity average molar mass according to the Mark-Houwink(1938) viscosity equation:

$$[\eta] = K\overline{M}_v^{\alpha} \tag{8.7}$$

where K and α are the Mark-Houwink parameters. These constants are unique for each combination of polymer and solvent and can be found tabulated in the appropriate reference literature. The Mark Houwink parameters given are in most cases based on samples with a narrow moral mass distribution. If eq. (8.7) is used for a polymer sample with a broad molar mass distribution, the molar mass value obtained is indeed the viscosity average.

All these averages are equal only for a perfectly monodisperse polymer. In all other cases, the averages are different: $M_n < M_v < M_w < M_z$. The viscosity average is often relatively close to the weight average.

1. New words

rheological[ˌriəˈlɑdʒikəl] *adj.* 流变学的,液流学的

oligomer[əˈligəmə] *n.* 低聚物,低聚体

viscometry[visˈkɔmitri] *n.* 黏度测定法

viscometer[visˈkɔmitə] *n.* 黏度计

2. Phrases and expressions

size exclusion chromatography 体积排阻色谱

intrinsic viscosity 特性黏度

3. Notes to the text

① Low molar mass substances(oligomers) show a strong increase in melting point with increasing molar mass, whereas a constant melting point is approached in the polymer molar mass range. 低摩尔质量物质(低聚物)的熔点随摩尔质量的增加而显著增加，而在聚合物摩尔质量范围内则接近恒定的熔点。

4. Exercise

(1) Describe the number average molar mass, weight average molar mass, and the Z average molar mass respectively and the differences among all these averages.

(2) Demonstrate Mark-Houwink viscosity equation.

Reading Material

Characterization of Polymer Molecular Weight

With the exception of a few naturally occurring polymers, all polymers consist of molecules with a distribution of chain lengths. It is therefore necessary to characterize the entire distribution quantitatively, or at least to define and measure average chain lengths or molecular weights for these materials, because many important properties of the polymer depend on these quantities①. Extensive reviews are available concerning the effects of molecular weight and molecular weight distribution on the mechanical properties of polymers.

The concept of an average molecular weight causes some initial difficulty because we're used to thinking in terms of ordinary low molecular weight compounds in which the molecules are identical and there is a single, well-defined molecular weight for the compound. Where the molecules in a sample vary in size, however, the results depend on how you count.

In the case of pure, low molecular weight compounds, the molecular weight is defined as

$$M = \frac{W}{N} \tag{8.8}$$

where W = total sample weight

N = number of moles in the sample

Where a distribution of molecular weights exists, a number-average molecular weight $\overline{M_n}$ may be defined in an analogous fashion to (8.8):

$$\overline{M_n} = \frac{W}{N} = \frac{\sum_{x=1}^{\infty} n_x M_x}{\sum_{x=1}^{\infty} n_x} = \frac{n_1 M_1}{\sum n_x} + \frac{n_2 M_2}{\sum n_x} + \cdots = \sum_{x=1}^{\infty} \left(\frac{n_x}{N}\right) M_x \tag{8.9}$$

Where W = total sample weight $= \sum_{x=1}^{\infty} w_x = \sum_{x=1}^{\infty} n_x M_x$

w_x = total weight of x-mer

N = total number of moles in the sample(of all size) = $\sum_{x=1}^{\infty} n_x$

n_x = number of moles of x-mer

M_x = molecular weight of x-mer

(n_x/N) = mole fraction of x-mer

Any analytical technique that determines the number of moles present in a sample of known weight, regardless of their size, will give the number-average molecular weight.

Rather than count the number of molecules of each size present in a sample, it is possible to define an average in terms of the *weights* of molecules present at each size level. This is the weight-average molecular weight $\overline{M}_w$②.

A typical synthetic polymer might consist of a mixture of molecules with degrees of polymerization x ranging from one to perhaps millions. The complete molecular weight distribution specifies the mole(number) or mass(weight) fraction of molecules at each size level in a sample.

Distributions are often presented in the form of a plot of mole(n_x/N) or mass(w_x/W) faction of x-met vs. either x or M_x. Since x and M_x differ by a constant factor, the molecular weight of the repeating unit m, it makes little difference which is used. Because x can assume only integral values, a true distribution must consist of a series of spikes, one at each integral value of x, or separated by m molecular weight units if plotted against M_x. The height of the spike represents the mole or mass fraction of that particular x-met. No analytical technique is capable of resolving the individual x-mers, so distribution are drawn(and represented mathematically) as continuous curves, the locus of the spike tops③.

1. New words

quantitatively['kwɔntitətivli] *adv.* 定量地

quantity['kwɔntəti] *n.* 数量

initial[i'niʃəl] *adj.* 初步,最初

identical[ai'dentikəl] *adj.* 相同的,一致的

analogous[ə'næləgəs] *adj.* 类似的,相似的

synthetic[sin'θetik] *adj.* 合成的,人造的

polymerization[ˌpɔliməraiˈzeiʃən] *n.* 聚合,聚合反应

specifies['spiːʃiːz] *n.* 种类,类型

integral['intigrəl] *adj.* 完整的,整体的

spike[spaik] *n.* 峰,峰形

plot[plɔt] *vt.* 策划,绘图

locus['ləukəs] *n.* 轨迹

2. Phrases and expressions

average molecular weight 平均相对分子质量

ordinary low molecular weight compounds 普通低分子化合物

in the case of 就……来说,关于

number-average molecular weight 数均相对分子质量

3. Notes to the text

① It is therefore necessary to characterize the entire distribution quantitatively, or at least to define and measure average chain lengths or molecular weights for these materials, because many important properties of the polymer depend on these quantities. 由于聚合物很多重要的性质取决于平均链段长度或平均相对分子质量,因此定量地表征整个分布或者至少定义和测量这些量是有必要的。

② Rather than count the number of molecules of each size present in a sample, it is possible to define an average in terms of the *weights* of molecules present at each size level. This is the weight-average molecular weight $\overline{M}_w$. 重均相对分子质量可以被定义为任意尺寸的相对分子质量的一个平均值而不必去数一个样品中每种尺寸的分子的数目。

③ No analytical technique is capable of resolving the individual *x*-mers, so distribution are drawn (and represented mathematically) as continuous curves, the locus of the spike tops. 没有任何分析技术能够解决个别的任一组分,因此分布被描述为(并在数学上表示为)连续的曲线,轨迹的最高点是峰顶。

Lesson 9 Polymer Solubility and Solutions

The thermodynamics and statistics of polymer solutions is an interesting and important branch of physical chemistry, and is the subject of many good books and large sections of books in itself. It is far beyond the scope of this chapter to attempt to cover the subject in detail. Instead, we will concentrate on topics of practical interest and try to indicate, at least qualitatively, their fundamental bases. Three factors are of general interest:

(1) What solvents will dissolve what polymers?

(2) How do the interactions between a polymer and a solvent influence the properties of the solution?

(3) To what applications do the interesting properties of polymer solutions lead?

Figure 9.1 shows schematically a phase diagram for a typical polymer-solvent system, plotting temperature vs. the fraction polymer in the system. At low temperatures, a two-phase system is formed. The dotted tie lines connect the compositions of phases in equilibrium, a solvent-rich (dilute-solution) phase on the left and a polymer-rich (swollen-polymer or gel) phase on the right. As the temperature is raised, the compositions of the phases become more nearly alike, until at the upper critical solution temperature (UCST) they are identical. Above the UCST, the system forms homogeneous (single-phase) solutions across the entire composition range. The location of the phase boundary depends on the molecular weight of the polymer and the interaction between the polymer and solvent.

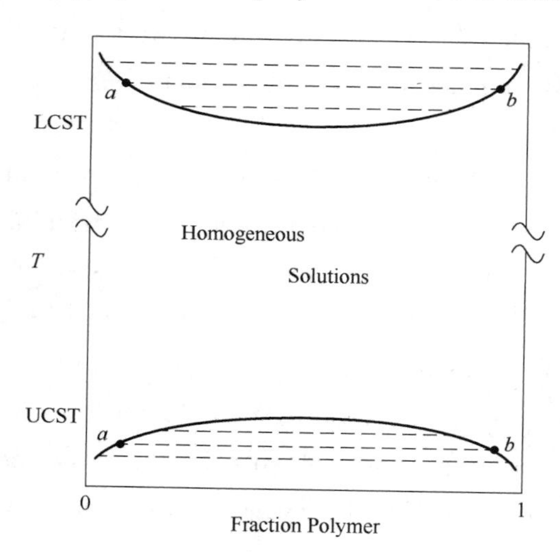

Figure 9.1 Schematic phase diagram for polymer-solvent system: *a*—dilute solution phase; *b*—swollen polymer or "gel" phase. UCST, upper critical solution temperature; LCST, lower critical solution temperature

In recent years, a number of systems have been examined that also exhibit a lower critical solution temperature (LCST), as shown in Figure 9.1. (One might question the nomenclature that puts the LCST above the UCST, but that's the way it is). LCSTs are more difficult to observe experimentally because they often lie well above the normal boiling points of the solvents.

When we talk about a polymer being soluble in particular solvent, we generally mean that the system lies between its LCST and UCST; That is, it forms homogeneous solutions over the entire composition range①. Keep in mind, however, that homogeneous solutions can still be formed toward the extremes of the composition range below the UCST and above the LCST.

Let's begin by listing some general qualitative observations on the dissolution of Polymers:

(1) Like dissolves like; that is, polar solvents will tend to dissolve polar polymers and nonpolar solvents will tend to dissolve nonpolar polymers. The chemical similarity of a polymer and a solvent is a fair indication of solubility②; for example, polyvinyl alcohol, $\left[\begin{matrix} \mathrm{H} & \mathrm{H} \\ | & | \\ \mathrm{C}- & \mathrm{C} \\ | & | \\ \mathrm{H} & \mathrm{O-H} \end{matrix}\right]_x$, will dissolve in water, H—O—H, and polystyrene, $\left[\begin{matrix} \mathrm{H} & \mathrm{H} \\ | & | \\ \mathrm{C}- & \mathrm{C} \\ | & | \\ \mathrm{H} & \Phi \end{matrix}\right]_x$, in toluene, $\Phi-CH_3$, but toluene won't dissolve polyvinyl alcohol and water won't dissolve polystyrene (which is good news for those of us who drink coffee out of foamed polystyrene cups).

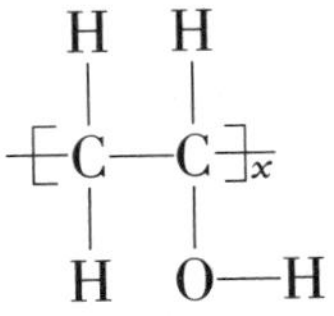

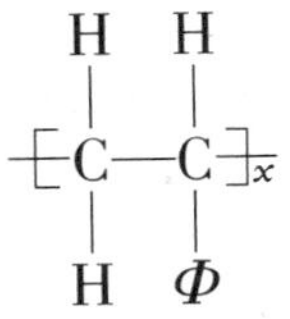

(2) In a given solvent at a particular temperature, the solubility of a polymer will decrease with increasing molecular weight.

(3) a. Crosslinking eliminates solubility.

b. Crystallinity, in general, acts like crosslinking, but it is possible in some cases to find solvents strong enough to overcome the crystalline bonding forces and dissolve the polymer③. Heating the polymer toward its crystalline melting point allows its solubility in appropriate solvents. For example, nothing dissolves polyethylene at room temperature. At 100℃, however, it will dissolve in a variety of aliphatic, aromatic, and chlorinated hydrocarbons.

(4) The rate of polymer solubility decreases with increasing molecular weight. For reasonably high molecular weight polymers, it can be orders of magnitude slower than that for nonpolymeric solutes.

It is important to note here that items 1, 2, and 3 are equilibrium phenomena and are therefore describable thermodynamically (at least in principle), while item 4 is a rate phenomenon and is governed by the rates of diffusion of polymer and solvent.

1. New words

solubility[ˌsɔlju'biliti] *n.* 溶解性

solution[sə'lu:ʃən] *n.* 溶液

thermodynamics[ˌθə: məudai'næmiks] *n.* 热力学

scope[skəup] *n.* 范围，机会，广度，眼界

dissolve[diəzɔv] *vi.* 溶解

qualitatively['kwɔlitətiv] *adv.* 定性

fraction['frækʃən] *n.* 分散

dotted['dɔtid] *adj.* 点线的
phase[feiz] *n.* 阶段
location[ləu'keitʃən] *n.* 定位,地点,位置
nomenclature[nəu'menklətʃə] *n.* 命名法,命名原则
soluble['sɔljubl] *adj.* 溶解的,可溶解的
homogeneous[ˌhɔməu'dʒiːnjəs] *adj.* 均匀的,均一的
dissolution[disə'ljuːʃən] *n.* 溶解
similarity[ˌsimi'læriti] *n.* 相似,类似,相似性,像
toluene['tɔljuiːn] *n.* 甲苯
foam[fəum] *v.* 发泡;*n.* 泡沫
aliphatic[ˌæli'fætik] *adj.* 脂肪族的
aromatic[ˌærəu'mætik] *adj.* 芳香的,醇香的,芬芳的
chlorinated['klɔːrineitid] *adj.* 氯化的
magnitude['mægnitjuːd] *n.* 大小,尺寸
solute['sɔljuːt] *n.* 溶质
diffusion[di'fjuːʒən] *n.* 扩散,漫射

2. Phrases and expressions

thermodynamics and statistics 热力学与统计学
concentrate on 将……集中于……
fundamental bases 基本依据
plotting temperature 策划温度
in equilibrium 平衡
upper critical solution temperature(UCST) 最高临界溶解(相容)温度
lower critical solution temperature(LCST) 最低临界溶解(相容)温度
polar solvents 极性溶剂
nonpolar solvents 非极性溶剂
polyvinyl alcohol 聚乙烯醇
the crystalline bonding forces 结晶黏结力
nonpolymeric solutes 非聚合物溶质

3. Notes to the text

① When we talk about a polymer being soluble in particular solvent, we generally mean that the system lies between its LCST and UCST; That is, it forms homogeneous solutions over the entire composition range. 当我们说某种聚合物在特定的溶剂中溶解时,我们一般是说这个系统位于 LCST 和 UCST 之间。也就是说,在整个混合物范围内形成了均匀的溶液。

② Like dissolves like; That is, polar solvents will tend to dissolve polar polymers and nonpolar solvents will tend to dissolve nonpolar polymers. The chemical similarity of a polymer and a solvent is a fair indication of solubility; 所谓相似相容也就是说极性溶剂溶解极性溶质,非极性溶剂溶解非极性溶质。聚合物和溶剂的化学相似性是溶解性的一个直接表征。

③ Crystallinity, in general, acts like crosslinking, but it is possible in some cases to find solvents strong enough to overcome the crystalline bonding forces and dissolve the polymer. 一般情况下,结晶行为类似交联,但是它在某些情况下可能有很强的溶剂来克服结晶黏结力并溶解聚合物。

4. Exercises

(1) What does the location of the phase boundary depend on in Figure 9. 1?

(2) Sketch qualitatively how you would expect the molecular weight of the polymer to affect the ternary phase diagram in Figure 9. 1.

(3) Why is it good news for those of us who drink coffee out of foamed polystyrene cups?

(4) Polyethylene will dissolve in aliphatic at ________.

a. 25℃   b. 100℃   c. −10℃   d. 55℃

Reading Material

The Thermodynamic Basis of Polymer Solubility and Solubility Parameter

The Thermodynamic Basis of Polymer Solubility

"To dissolve or not to dissolve. That is the question." (with apologies to W. S.) . The answer is determined by the sign of the Gibbs free energy. Consider the process of mixing pure polymer and pure solvent (state 1) at constant pressure and temperature to form a solution (state 2):

$$\Delta G = \Delta H - T\Delta S \tag{9.1}$$

where ΔG = the change in Gibbs free energy

ΔH = the change in enthalpy

T = the absolute temperature

ΔS = the change in entropy

Only if ΔG is negative will the solution process be thermodynamically feasible. The absolute temperature must be positive, and the change in entropy for a solution process is generally positive, because in a solution, the molecules are in a more random state in the solid[①] (this might not always be the case with lyotropic liquid-crystal materials) . The positive product is preceded by a negative sign. Thus, the third ($-T\Delta S$) term in (9. 1) favors solubility. The change in enthalpy may be either positive or negative. A positive ΔH means the solvent and polymer "prefer their own company," that is, the pure materials are in a higher energy state, while a negative ΔH indicates that the solution is the lower energy state. If the latter obtains, solution is assured. Negative ΔH's usually arise where specific interactions such as hydrogen bonds are formed between the solvent and polymer molecules. But, if ΔH is positive, then $\Delta H < T\Delta S$, the polymer is to be soluble.

One of the things that makes polymers unusual is that the entropy change in forming a polymer solution is generally much smaller than that which occurs on dissolution of equivalent masses or volumes of low molecular weight solutes②. The reasons for this are illustrated qualitatively on a two-dimensional lattice model in Figure 9. 2. With the low molecular weight solute, the solute molecules may be distributed randomly throughout the lattice, the only restriction being that a lattice site cannot be occupied simultaneously by two (or more) molecules. This gives rise to a large number of configurational possibilities, that is, a high entropy. In the polymer solution, however, each chain segment is confined to a lattice site adjacent to the next chain segment, greatly reducing the configurational possibilities. Note also that for a given number of chain segments (equivalent masses or volumes of polymer) the more chains they are split up into, that is, the lower their molecular weight, the higher will be their entropy upon solution③. This explains directly observation 2, the decrease in solubility with molecular weight. But in general, for high molecular weight polymers, because the $T\Delta S$ term is so small, if ΔH is positive then it must be even smaller if the polymer is to be soluble. So in the absence of specific interactions, predicting polymer solubility largely boils down to minimizing ΔH.

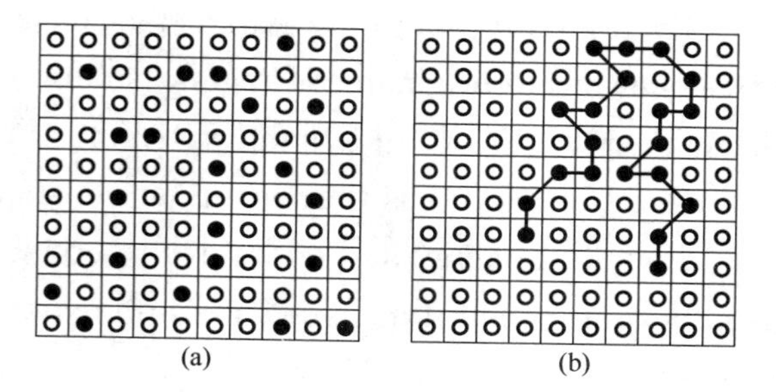

Figure 9. 2 Lattice model of solubility: (a) low molecular weight solute (b) polymeric solute
○ Solvent, ● Solute

The Solubility Parameter

How can ΔH be estimated? Well, for the formation of regular solutions (those in which solute and solvent do not form specific interactions), the change in internal energy per unit volume of solution is given by

$$\Delta H \approx \Delta E = \Phi_1 \Phi_2 (\delta_1 - \delta_2)^2 \text{cal/cm}^3 \tag{9.2}$$

where ΔE = the change in internal energy per unit volume of solution

Φ_i = volume fractions

δ_i = solubility parameters

The subscripts 1 and 2 usually (but not always!) refer to solvent and solute (Polymer), respectively. The solubility parameter is defined as follows:

$$\delta = (CED)^{1/2} = (\Delta E_V / V)^{1/2} \tag{9.3}$$

where CED = cohesive energy density, a measure of the strength of the intermolecular forces holding the molecules together in the liquid state

ΔE_V = molar change in internal energy on vaporization

V = molar volume of liquid.

Traditionally, solubility parameters have been given in $(cal/cm^3)^{1/2}$ = hildebrands (in honor of the originator of regular solution theory), but they are now more commonly listed in $(MPa)^{1/2}$ [1 hildebrand = 0.4889 $(MPa)^{1/2}$].

Now, for a process that occurs at constant volume and constant pressure, the changes in internal energy and enthalpy are equal. Since the change in volume on solution is usually quite small, this is a good approximation for the dissolution of polymers under most conditions, so (9.2) provides a means of estimating enthalpies of solution if the solubility parameters of the polymer and solvent are known.

Note that regardless of the magnitudes of δ_1 and δ_2 (they must be positive), the predicted ΔH is always positive, because (9.2) applies only in the absence of the specific interactions that lead to negative ΔH's. Inspection of (9.2) also reveals that ΔH is minimized, and the tendency toward solubility is therefore maximized by matching the solubility parameters as closely as possible as a very rough rule-of-thumb (or heuristic principle, if you prefer),

$$|\delta_1 - \delta_2| < 1(cal/cm^3)^{1/2} \quad \text{for solubility} \tag{9.4}$$

1. New words

enthalpy['enθælpi] *n.* 焓
entropy['entrəpi] *n.* 熵
lyotropic[lai'ɔtrɔpik] *adj.* 溶致的,易溶的
equivalent[i'kwivələnt] *adj.* 相当的,相等的
restriction[ris'trikʃən] *n.* 限制,限定
simultaneously[siməl'teiniəsli] *adv.* 同时发生,一起,同时,同时存在
configurational[kənˌfigju'reiʃən] *adj.* 构造的
confined[kən'faind] *adj.* 限于,被限制的,狭窄的
subscript[səb'skrip] *n.* 下标
intermolecular[ˌintə(:)mə'lekjulə] *adj.* 分子间的,存在(或作用)于分子间的
vaporization[ˌveipərai'zeiʃən] *n.* 汽化
originator[ə'ridʒəneitə] *n.* 起源,起因,来源
magnitude['mægnitju:d] *n.* 巨大,重要性,程度
inspection[in'spekʃən] *n.* 检查
minimize['minimaiz] *vt.* &*vi.* 最小化
maximize['mæksimaiz] *vt.* 最大化

2. Phrases and expressions

the Gibbs free energy　吉布斯自由能
lyotropic liquid-crystal　溶致液晶
two-dimensional lattice model　二维格子模型
give rise to　引起,导致
equivalent masses　等效质量
split up into　分割成……
boils down to　得出……的结论,归结为……
cohesive energy density　内聚能密度
rule-of-thumb　经验法则
heuristic principle　启发性原则
hildebrands　希耳德布兰特,hildebrand function(theory)希耳德布兰特函数(理论)

3. Notes to the text

① Only if ΔG is negative will the solution process be thermodynamically feasible. The absolute temperature must be positive, and the change in entropy for a solution process is generally positive, because in a solution, the molecules are in a more random state in the solid 只有吉布斯自由能为负值时,热力学的溶解过程才能进行。绝对温度必须是正值,溶解过程中熵的变化一般是增大的,因为分子在溶液中比在固体中更易处于一种无规状态。

② One of the things that makes polymers unusual is that the entropy change in forming a polymer solution is generally much smaller than that which occurs on dissolution of equivalent masses or volumes of low molecular weight solutes. 聚合物的特性之一是在形成聚合物溶液的过程中熵的变化一般比等量或等体积的低相对分子质量溶质的溶解过程中的要小得多。

③ Note also that for a given number of chain segments(equivalent masses or volumes of polymer) the more chains they are split up into, that is, the lower their molecular weight, the higher will be their entropy upon solution. 还要注意,对于给定数量的链段(等量或等体积聚合物),它们分成的链数越多,即相对分子质量越小,溶液中的熵越大。

Lesson 10 Transitions in Polymers

The Glass Transition

It has long been known that amorphous polymers can exhibit two distinctly different types of mechanical behavior. Some, like polymethyl methacrylate (Lucite, Plexiglas) and polystyrene, are hard, rigid, glassy plastics at room temperature, while others, for example, polybutadiene, polyethyl acrylate, and polyisoprene, are soft, flexible rubbery materials. If, however, polystyrene and polymethyl methacrylate are heated to about 125℃, they exhibit typical rubbery properties, and when a rubber ball is cooled in liquid nitrogen, it becomes rigid and glassy, and shatters when an attempt is made to bounce it. So, there is some temperature, or narrow range of temperatures, below which an amorphous polymer is in a glassy state, and above which it is rubbery. This temperature is known as the glass transition temperature T_g. The glass transition temperature is a property of the polymer, and whether the polymer has glassy or rubbery properties depends on whether its application temperature is above or below its glass transition temperature ①.

Molecular Motions in an Amorphous Polymer

To understand the molecular basis for the glass transition, the various molecular motions occurring in an amorphous polymer mass may be broken into four categories:

(1) Translational motion of entire molecules, which permits flow.

(2) Cooperative wriggling and jumping of segments of molecules approximately 40 to 50 carbon atoms in length, permitting the flexing and uncoiling which lead to elasticity.

(3) Motions of a few atoms along the main chain (five or six, or so) or of side groups on the main chains.

(4) Vibrations of atoms about equilibrium positions, as occurs in crystal lattices, except that the atomic centers are not in a regular arrangement in an amorphous polymer.

Motions 1-4 above are arranged in order of decreasing activation energy; That is, smaller amounts of thermal energy kT are required to produce them. The glass transition temperature is thought to be that temperature at which motions 1 and 2 are pretty much "frozen out," and there is only sufficient energy available for motions of types 3 and 4 ②. Of course, not all molecules possess the same energies at a given temperature. The molecular energies follow a Boatman distribution, and even below T_g, there will be occasional type 2 and even type 1 motions, which can manifest themselves over extremely long periods of time.

Determination of T_g

How is the glass transition studied? A common method is to observe the variation of some thermodynamic property with T, for example, the specific volume, as shown in Figure 10. 1. Note that the slope of the V vs. T plot increases above the glass transition temperature ③.

Factors That Influence T_g

In general, the glass transition temperature depends on five factors.

a. The free volume of the polymer V_f;

b. The attractive between the molecules;

c. The internal mobility of the chains;

d. The stiffness of the chains;

e. The chain length.

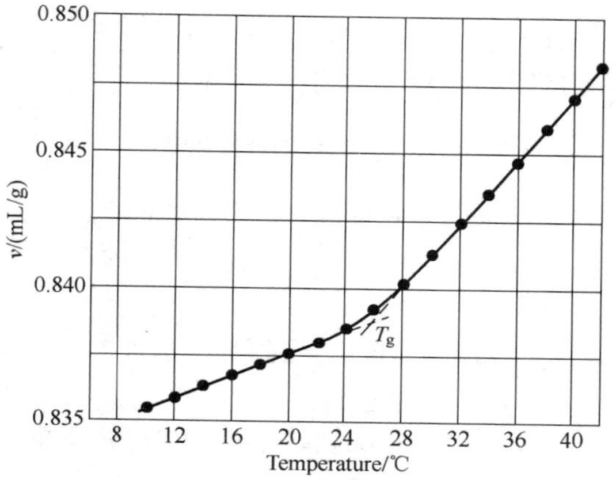

Figure 10. 1 Specific volume vs. temperature for polyvinyl acetate

Other Transitions

Transitions other than T_g and Tm are sometimes observed in polymers. Some polymers possess more than one crystal form, so there will be an equilibrium temperature of transition from one to another. Similarly, second-order transitions below T_g occur in some materials (T_g is then termed the α transition, the next lower the β, etc.) ④. These are attributed to motions of groups of atoms smaller than those necessary to produce T_g. These transitions may strongly influence properties. For example, tough amorphous plastics (e. g. polycarbonate) have such a transition well below room temperature, while brittle amorphous plastics (e. g. polystyrene and polymethyl methacrylate) do not.

The existence of another transition above T_g has been claimed, but is still the subject of considerable controversy. This T_{11} (liquid-liquid transition) presumably represents the boundary between type 1 and type 2 motions. It has been observed in a number of systems, and it has been suggested that $T_{11} \approx 1.2\ T_g$ for all polymers. For each article that reports T_{11}, however, it seems that there is another that claims that T_{11} results from impurities (traces of solvent or unreacted monomer) in the sample or is an artifact of experimental or data-analysis technique.

1. New words

polybutadiene[ˌpɔliˌbjuːtə'daIiːn]n. 聚丁二烯
polyisoprene[ˌpɔli'aisəupriːn]n. 聚异戊二烯
wriggle['rigl]vi. 蠕动,蜿蜒行进,扭动
nitrogen['naitrədʒən]n. 氮
elasticity[læsi'tisiti]n. 弹性,弹力
manifest['mænifest]vt. 表明,表现,证明
specific[spi'sifik]adj. 特定的
volume['vɔljuːm]n. 体积
slope[sləup]n. 斜率
controversy['kɔntrəvəːsi]n. 争议,论争,辩论
presumably[pri'zjuːməbəli]adv. 推测,大概
impurity[im'pjuəriti]n. 杂质
artifact['ɑːtifækt]n. 人造物品

2. Phrases and expressions

amorphous polymers　非晶态聚合物
polymethyl methacrylate　聚甲基丙烯酸甲酯
poly ethyl acrylate　聚丙烯酸乙酯
glassy state　玻璃态
glass transition temperature　玻璃化转变温度
translational motion　平移运动
cooperative wriggling　共同蠕动
specific volume　比体积
activation energy　活化能
frozen out　冻结
more than　不只,超过

3. Notes to the text

① The glass transition temperature is a property of the polymer, and whether the polymer has glassy or rubbery properties depends on whether its application temperature is above or below its glass transition temperature.　玻璃化转变温度是聚合物的一种性质,聚合物是玻璃态还是橡胶态取决于它的使用温度是高于还是低于玻璃化转变温度。

② The glass transition temperature is thought to be that temperature at which motions 1 and 2 are pretty much "frozen out," and there is only sufficient energy available for motions of types 3 and 4.　运动类型1和2都被"冻结",同时只有足够的能量用于运动类型3和4的温度被认为是玻璃化转变温度 。

③ A common method is to observe the variation of some thermodynamic property with T, for example, the specific volume, as shown in Figure 10. 1. Note that the slope of the V vs. T plot increases above the glass transition temperature.　一种常用的方法是观察随温度的变化而变化的某种热力学性质,例如图10.1中的具体比体积。要注意体积对温度曲线的斜率在玻璃化转变温度以上增大了。

④ Transitions other than T_g and Tm are sometimes observed in polymers. Some polymers possess more than one crystal form, so there will be an equilibrium temperature of transition from one to another. Similarly, second-order transitions below T_g occur in some materials(T_g is then termed the α transition, the next lower the β, etc.).　除了玻璃化

转变温度与熔点外,聚合物中还可以观察到其他的转变。一些聚合物不只有一种结晶形式,所以从一种形式到另一种形式的转变就存在一个转变平衡温度。同样,在一些材料中在玻璃化温度以下时发生二级转变。(此时的玻璃化转变温度称为一级转变 α 转变,二级转变 β 转变等)。

4. Exercises

(1) Which rubbery material isn't soft, flexible at room temperature?

a. Polybutadiene.
b. Polyethyl acrylate.
c. Polyisoprene.
d. Polymethyl methacrylate.

(2) In question 1, can you explain why it isn't soft, flexible at room temperature?

(3) Amorphous polymers are generally preferred to crystalline polymers. Why?

(4) The existence of another transition above Tg has been claimed, but is still the subject of considerable controversy. Why?

Reading Material

Orientation and Drawing of Polymer

When a polymer mass is crystallized in the absence of external forces, there is no preferred direction in the specimen along which the polymer chains lie①. If such an unoriented crystalline polymer is subjected to an external stress, it undergoes a rearrangement of the crystalline material. Changes in the *x*-ray diffraction pattern suggest that the polymer chains align in the direction of the applied stress. At the same time the physical properties of the sample change markedly.

When the orientation process is carried out below T_m but above the glass transition temperature T_g, as when an unoriented fiber is stretched rapidly, the sample does not become gradually thinner but suddenly becomes thinner at one point, in a process known as "necking down". As the stretching is continued, the thin or drawn section increases in length at the expense of the undrawn portion of the sample②. The diameters of the drawn and undrawn portions remain about the same throughout the process. The draw ratio, or ratio of the length of the drawn fiber to that of the undrawn, is about 4 or 5 to 1 for a number of polymers, including branched polyethylene, polyesters, and polyamides, but is much higher (10 to 1 or more) in linear polyethylene.

In general, the degree of crystallinity in the specimen does not change greatly during drawing if crystallinity was previously well developed③. If the undrawn polymer was amorphous or only partially crystallized, crystallinity is likely to increase during cold drawing.

Observations in the electron microscope indicate that, when polymer single crystals are stressed, fibrils 5.0 ~ 10.0nm in diameter are drawn across the break[④]. These fibrils must contain many molecules; smaller fibrils have not been resolved in the microscope. The fibrils appear to come from individual lamellae, and the sharp discontinuity between the drawn and the undrawn material suggests that the molecules are unfolding. In some cases the fibrils appear to have a periodicity of 10.0 ~ 40.0nm along their length; a similar period is observed by low-angle *x*-ray diffraction.

Examination of melt or solution cast specimens shows that the lamellae tend to rotate toward the draw direction and on further elongation break up into micro-lamellae and finally into submicroscopic units. Since the elongation at break in cold drawn specimens is less than could be obtained by complete unfolding[⑤], it has been suggested that the smallest units are micro-crystallites in which the chains remain folded. Periodic spacing of these crystallites within the fibrils would account for their observed periodicity.

Spherulites tend to remain intact during the first stages of drawing, often elongating to markedly ellipsoidal shapes. Rupture of the sample usually occurs at spherulite boundaries. In some cases it is accompanied by the production of fibrous material.

The structure of fibers, whether formed by drawing or crystallized from an oriented melt, is still unsettled. X-ray evidence indicates that the molecules are aligned and shows that a long periodicity is present. Some sort of folded structure is regarded as possible, although the available evidence can also be explained in terms of alternating crystalline-amorphous regions within elementary fibrils consisting of aggregates of several hundred molecules. A nonfolded paracrystalline structure has also been postulated in which loops or isolated folds and chain ends form dislocations and lattice vacancies analogous to those found in metals.

Annealing of both drawn and undrawn material indicates a recrystallization process in which chain refolding takes place to give more ordered structures with higher melting points. In undrawn material, recrystallization is preceded by premelting (substantial loss of crystalline order), whereas in drawn material premelting is not observed. Hence it is concluded that melting is not a prerequisite to refolding of chains.

1. New words

unoriented[ʌn'ɔːrientid] *adj.* 未取向的
stretch[stretʃ] *vi.* & *vt.* 拉伸，伸展
fibril['faibril] *n.* 微纤
unfold[ʌn'fəuld] *vt.* & *vi.* 展开，打开
periodicity[piəriə'disiti] *n.* 周期性
lamellae[læ'meliː] *n.* 晶片
elongation[ˌilɔːŋ'geiʃən] *n.* 伸长率
spherulite['sferjulait] *n.* 球晶
intact[in'tækt] *adj.* 完整的，整体的

2. Phrases and expressions

necking down 成颈，缩颈
X-ray diffraction pattern X射线衍射图

draw ratio 拉伸比
cold drawing 冷拉伸
polymer single crystal 聚合物单晶
individual lamellae 个别晶片
low-angle x-ray diffraction 小角 X 射线衍射
ellipsoidal shape 椭圆形

3. Notes to the text

① When a polymer mass is crystallized in the absence of external forces, there is no preferred direction in the specimen along which the polymer chains lie. 聚合物材料在没有外力存在下结晶时，聚合物链在样品中没有择优的方向。

② As the stretching is continued, the thin or drawn section increases in length at the expense of the undrawn portion of the sample. 随着拉伸过程的延续，样品的未拉伸部分逐渐减少，而细的或被拉伸部分逐渐增长。

③ In general, the degree of crystallinity in the specimen does not change greatly during drawing if crystallinity was previously well developed. 一般来说，如果结晶性已经发展得很好时，拉伸过程中样品的结晶度不会有很大变化。

④ Observations in the electron microscope indicate that, when polymer single crystals are stressed, fibrils 5.0 ~ 10.0nm in diameter are drawn across the break. 通过电子显微镜观察会发现，当聚合物单晶被拉伸时在断裂处被拉出直径为 5.0 ~ 10.0nm 的微纤。

⑤ Since the elongation at break in cold drawn specimens is less than could be obtained by complete unfolding, … 因为冷拉样品的断裂伸长比完全展开所应有的断裂伸长小，……

PART 3 POLYMER SYNTHESIS

Lesson 11 Step-Growth (Condensation) Polymerization

Polymerization reactions may be divided into two categories according to the mechanism by which the chains grow. In step-growth polymerization, also known as condensation polymerization, chains of any lengths x and y combine to form longer chains:

$$x\text{-mer} + y\text{-mer} \rightarrow (x+y)\text{-mer} \quad \text{(step growth)} \tag{11.1a}$$

In chain-growth (addition) polymerization, a chain of any length x can only add a monomer molecule and continue its growth:

$$x\text{-mer} + \text{monomer} \rightarrow (x+1)\text{-mer} \quad \text{(chain growth)} \tag{11.1b}$$

By classifying the reactions according to either of the above mechanisms, we avoid any confusion. Step-growth polymerization is treated in this chapter and various types of chain growth are discussed in another two Chapters.

Regardless of the type of polymerization reaction, quantitative treatments are usually based on the assumption that the reactivity of the functional group at a chain end is independent of the length of the chain[①], x and y in (11.1a) and (11.1b) above. Experimentally, this is an excellent assumption for x's greater than about five or six. Since most polymers must develop x's on the order of a hundred or so to be of practical value, this assumption introduces little error while enormously simplifying the mathematical treatment. With these basic concepts in mind, we proceed to a more detailed and quantitative treatment of polymerization reactions.

Since a distribution of chain lengths always arises in these reactions, the average chain length is often of interest. Each molecule in the reaction mass has, on the average, one un-reacted A group. If N_0 is the original number of molecules present in the reaction mass, at the start of the reaction there are N_0 un-reacted A groups present. At some later time during the reaction there are N molecules, and therefore N un-reacted A groups present, so $N_0 - N$ of the A groups have reacted. Thus,

$$p = \frac{N_0 - N}{N_0} = \text{fraction of reacted A groups present} \tag{11.2}$$

Because the N_0 original monomer molecules are distributed among the N molecules present in the reaction mass, the average number of monomer residues per molecule $\overline{x_n}$ is given by

$$\overline{x}_n = \frac{N}{N_0} = \text{average number of monomer residues per molecule} \qquad (11.3)$$

Combining(11.2)and(11.3)gives

$$\overline{x}_n = \frac{1}{1-p} \quad (\text{Carothers' equation}) \qquad (11.4)$$

This rather simple conclusion was reached by W. H. Carothers, the discoverer of nylon and one of the founders of polymer science, in the 1930s, but it is often ignored to this day. Its importance becomes obvious when it is realized that typical linear polymers must have $\overline{x}_n$ values on the order of 100 to achieve useful mechanical properties. This requires a conversion of at least 99% , assuming difunctional monomers in perfect stoichiometric equivalence. Such high conversions are almost unheard of in most organic reactions, but are necessary to achieve high molecular weight condensation polymers.

Since all reactions are reversible, many polycondensations would reach equilibrium at low conversions if the molecule of condensation were not efficiently removed(e. g. , with heat and vacuum or by a second reaction) to drive the reaction to high conversions②.

1. New words

molecule['mɔliku:l]*n.* 分子

reaction[ri(:)'ækʃən]*n.* 反应

enormously[i'nɔ:məsli]*adv.* 非常地，巨大地

fraction['frækʃən]*n.* 小部分，片段，分数

residues[ri'zidjuəm]*n.* 残余，剩余物，残数，滤渣，渣滓

stoichiometry[ˌstɔiki'ɔmitri]*n.* 化学计算(法)，化学计量学

equivalence[i'kwivələns]*n.* 等价，等值

2. Phrases and expressions

be divided into　分为……，归结为……

regardless of　不管，不顾

mathematical treatment　精确处理

quantitative treatment　定量处理

mechanical properties　力学性能，机械性能

3. Notes to the text

① Regardless of the type of polymerization reaction, quantitative treatments are usually based on the assumption that the reactivity of the functional group at a chain end is independent of the length of the chain.　无论聚合反应是何种类型，定量处理的基础一般被假定为分子链端官能团的活性与分子链的长短无关。

② Since all reactions are reversible, many polycondensations would reach equilibrium at low conversions if the molecule of condensation were not efficiently removed (e. g. , with heat and vacuum or by a second reaction) to drive the reaction to high conversions.　由于所有的反应都是可逆的，因此如果不能有效地去除浓缩产生的分子(例如用

加热和真空或二次反应的方法)来使反应达到高转化,很多缩合反应会在转化率很低的情况下就达到平衡。

4. Exercises

(1) Translate the text into Chinese.

(2) Describe the process of step-growth polymerization.

(3) What differences are there between step reaction polymerizations and condensation polymerizations?

(4) What are the quantitative treatments usually based on?

(5) Why should the molecule of condensation be efficiently removed?

Reading Material

Step Reaction Polymerization

Many simple organic reactions are known in which two molecules become joined, typical examples being the condensation of an acid with an alcohol to yield an ester and the similar reaction of an acid with an amine to yield an amide. Reactions of this type may be adapted to the formation of polymers by using molecules which are functionally capable of coupling indefinitely by condensation reactions, that is to say molecules having more than one carboxyl, amine or alcohol functio[①]. Difunctionality may be achieved by using a single monomer bearing two different functions as for example in the polymerization of a hydroxy acid to a polyester

$$n\mathrm{HO{-}(CH_2)_x{-}COOH} \rightarrow \mathrm{H[O{-}(CH_2)_x{-}\overset{\overset{\displaystyle O}{\|}}{C}]_n\,O{-}H} + (n-1)\mathrm{H_2O}$$

Alternatively the two functional groups may be present on different molecules as for example in the reaction of a diacid with diamine to yield a linear polyamide.

$$n\mathrm{HOOC{-}(CH_2)_4{-}COOH} + \mathrm{H_2N{-}(CH_2)_6{-}NH_2} \rightarrow$$

$$\mathrm{H{-}O[\overset{\overset{\displaystyle O}{\|}}{C}{-}(CH_2)_4{-}\overset{\overset{\displaystyle O}{\|}}{C}{-}\overset{\overset{\displaystyle H}{|}}{N}{-}(CH_2)_6{-}\overset{\overset{\displaystyle H}{|}}{N}]_n H} + (n-1)\mathrm{H_2O}$$

In this latter example the repeat unit of the polymer is produced from both starting materials-despite the fact that two monomers are used the polymer has only a single repeat unit (contained within the brackets in the formula) and is thus formally classified as a homopolymer[②]. The term copolymer is reserved for more complex cases in which the polymer contains more than one repeat unit.

Most of the coupling reactions used for polymer production are condensation reactions, in which the coupling is accompanied by the elimination of a small molecule such as water, hydrogen chloride or methanol. For this reason such reactions are frequently termed condensation polymerizations; It must be emphasized that this term refers to a specific polymerization mechanism in which growth occurs by a step-wise coupling of small molecules, irrespective of whether small molecules are eliminated or not③.

$$HO—R—OH + OCN—R'—NCO \rightarrow \left[O—R—O—\overset{\overset{O}{\|}}{C}—\overset{\overset{H}{|}}{N}—R'—\overset{\overset{H}{|}}{N}—\overset{\overset{O}{\|}}{C} \right]_n$$

Thus the reaction of a diol with a diisocyanate to produce a polyurethane proceed without elimination of any small molecule and is therefore not a condensation reaction. Nevertheless both the mechanism of polymerization and the structure of the final product are typical of condensation polymerizations and the reaction is normally considered as such. In many ways the description of these reactions as step reaction polymerizations is both preferable and more correct, but the term condensation is now firmly entrenched in the literature of polymer science. In recent years the concept of step reaction polymerization has been utilized to produce a very wide range of polymeric materials having useful thermal and mechanical characteristics. Although there is usually apparent formal elimination of a small molecule during the growth processes, the reaction mechanisms are frequently very much more complex.

1. New words

amine['æmi:n] *n.* 胺
amide['æmaid] *n.* 酰胺
couple['kʌpl] *n.* 连接,配合
carboxyl[ka:'bɔksil] *n.* 羧基
diamine[daiə'mi:n] *n.* 二胺化合物
polyamide[ˌpɔli'æmaid] *n.* 聚酰胺
homopolymer['həumə'pɔlimə] *n.* 均聚物
elimination[iˌlimi'neiʃən] *n.* 消除
stepwise['stepwaiz] *adj.* 逐步的
diol['dai'ɔl] *n.* 二醇
diisocyanate[daiˌaisə'saiəneit] *n.* 二异氰酸酯
polyurethane[ˌpɔli'jurəiθein] *n.* 聚氨酯
entrench[in'trentʃ] *v.* 牢固树立,确定
characteristic [ˌkæriktə'ristik] *n.* 特征; *adj.* 特有的,典型的,特性

2. Phrases and expressions

that is to say 即,也就是说,换句话说,更确切地说
difunctionality 二官能性
hydroxy acid 羟基酸
hydrogen chloride 氯化氢
step-wise 逐步

3. Notes to the text

① Reactions of this type may be adapted to the formation of polymers by using molecules which are functionally capable of coupling indefinitely by condensation reactions, that is to say molecules have more than one carboxyl, amine or alcohol function.
这种类型的反应可用来生成聚合物,如果所用的分子从功能上看,能够通过缩合反应无限连

接起来,即分子是带有多于一个的羧基、胺基或羟基。

② In this latter example the repeat unit of the polymer is produced from both starting materials-despite the fact that two monomers are used the polymer has only a single repeat unit and is thus formally classified as a homopolymer.　在后一个例子中,聚合物的重复单元是由两种起始物质形成的,尽管用两种单体,而实际上聚合物仅仅有一个重复单元。因此,形式上把它归为均聚物。

③ It must be emphasized that this term refers to a specific polymerization mechanism in which growth occurs by a step-wise coupling of small molecules, irrespective of whether small molecules are eliminated or not.　必须强调,这个术语是专指一个特殊的聚合机理,按照这个机理,不管小分子是否析出,增长的发生是通过小分子逐步结合的。

Lesson 12 Free-Radical Addition (Chain-Growth) Polymerization

One of the most important types of addition polymerization is initiated by the action of free radicals, electrically neutral species with an unshared electron[①]. In the developments to follow, a dot (·) will represent a single electron. The single bond, a pair of shared electrons, will be denoted by a double dot (:) or, where it is not necessary to indicate electronic configurations, by the usual sign (-). A double bond, two shared electron pairs, is indicated by : : or =.

Free radicals for the initiation of addition polymerization are usually generated by the thermal decomposition of organic peroxides or azo compounds. Two common examples are

Benzoyl peroxide

$$\Phi-\overset{\overset{\displaystyle O}{\|}}{C}-O:O-\overset{\overset{\displaystyle O}{\|}}{C}-\Phi \rightarrow 2\Phi-\overset{\overset{\displaystyle O}{\|}}{C}-O\cdot \rightarrow 2\Phi\cdot + CO_2$$

Azobisisobutyronitrile

$$(CH_3)_2-\underset{\underset{\displaystyle C\equiv N}{|}}{C}:N{=}N:\underset{\underset{\displaystyle C\equiv N}{|}}{C}-(CH_3)_2 \rightarrow 2(CH_3)_2-\underset{\underset{\displaystyle C\equiv N}{|}}{C}\cdot + N_2$$

The initiator molecule, represented by I, undergoes a first-order decomposition with a rate constant k_d to give two free radicals, R · :

$$I \xrightarrow{k_d} 2R \qquad \text{Decomposition} \qquad (12.1)$$

The radical then adds a monomer by grabbing an electron from the electron-rich double bond, forming a single bond with the monomer, but leaving an unshared electron at the other end:

$$R\cdot + \underset{\underset{\displaystyle H}{|}}{\overset{\overset{\displaystyle H}{|}}{C}}::\underset{\underset{\displaystyle X}{|}}{\overset{\overset{\displaystyle H}{|}}{C}} \rightarrow R:\underset{\underset{\displaystyle H}{|}}{\overset{\overset{\displaystyle H}{|}}{C}}:\underset{\underset{\displaystyle X}{|}}{\overset{\overset{\displaystyle H}{|}}{C}}\cdot$$

This may be abbreviated by

$$R\cdot + M \xrightarrow{k_a} P_1 \qquad \text{Addition} \qquad (12.2)$$

where P_1 · represents a growing polymer chain with 1 repeating unit. Note that the product of the addition reaction is still a free radical; it proceeds to propagate the chain by adding another monomer unit:

$$P_1\cdot + M \xrightarrow{k_p} P_2\cdot$$

again maintaining the unshared electron at the chain end, which adds another monomer unit:

$$P_2\cdot + M \xrightarrow{k_p} P_3\cdot$$

and so on. In general, the propagation reaction is written as

$$P_x\cdot + M \xrightarrow{k_p} P_{(x+1)}\cdot \quad \text{Propagation} \qquad (12.3)$$

We have again assumed that reactivity is independent of chain length by using the same k_p for each propagation step.

Growing chains can be terminated in one of two ways. Two can bump together and stick, their unshared electrons combining to form a single bond between them (combination):

$$\begin{array}{ccccccccccccccc} & H & & H & & H & & H & & H & & H & H & & H \\ & | & & | & & | & & | & & | & & | & | & & | \\ \sim & C & — & C\cdot & + & \cdot C & — & C & \sim\rightarrow\sim & C & — & C: & C & — & C \\ & | & & | & & | & & | & & | & & | & | & & | \\ & H & & X & & X & & H & & H & & X & X & & H \end{array}$$

$$P_x\cdot + P_y\cdot \xrightarrow{k_{tc}} \overline{P}_{(x+y)} \quad \text{Combination} \qquad (12.4a)$$

Where $P_{(x+y)}$ is a dead polymer chain of $(x+y)$ repeating units. Or, one can abstract a proton from the penultimate carbon of the other (*disproportionation*):

$$\begin{array}{cccccccccccccccc} & H & & H & & H & & H & & H & & H & & H & & H \\ & | & & | & & | & & | & & | & & | & & | & & | \\ \sim— & C & — & C\cdot & + & \cdot C & — & C & \sim\rightarrow\sim & C & = & C & + H: & C & — & C\sim \\ & | & & | & & | & & | & & & & | & & | & & | \\ & H & & X & & X & & H & & & & X & & X & & H \end{array}$$

$$P_x\cdot + P_y\cdot \xrightarrow{k_{td}} P_x - P_y \quad \text{Disproportion} \qquad (12.4b)$$

The relative proportion of each termination mode depends on the particular polymer and the reaction temperature, but in most cases, one or the other predominates②.

1. New words

initiate[iˈniʃieit] *vt.* 引发,开始,发动,传授;*v.* 开始,发起

dot[dɔt] *n.* 点,圆点

decomposition[ˌdiːkɔmpəˈziʃən] *n.* 分解,腐烂

azo[ˈæzəu] *adj.* 含氮的

azobisisobutyronitrile (AIBN) [ˌæzəubis ˈaisouˌbjuːtiˈrɔnaitrail] *n.* 偶氮二异丁腈

propagate[ˈprɔpəgeit] *v.* 繁殖,传播,宣传

penultimate[piˈnʌltimit] *adj.* 倒数第二个音节的

disproportionation[disprəpɔːʃəˈneiʃən] *n.* 不均,不对称,氢原子转形,歧化

predominate[priˈdɔmineit] *vt.* 起主导作用,居支配地位

2. Phrases and expressions

single bond 单键

free radicals 自由基

organic peroxide 有机过氧化物

be abbreviated by 简化为,缩短为

3. Notes to the text

① One of the most important types of addition polymerization is initiated by the action of free radicals, electrically neutral species with an unshared electron. 加成聚合反应中最重要的一种类型是由自由基引发的，这些自由基电子呈电中性并带有一个未配对的电子。

② The relative proportion of each termination mode depends on the particular polymer and the reaction temperature, but in most cases, one or the other predominates. 每种终止形式之间的比例取决于聚合物本身以及反应温度，但是在大多数情况下是一种或是另一种起主导作用。

4. Exercises

(1) What initiates one of the most important types of addition polymerization? And, what are their characteristics?

(2) How are the single bond and the double bond indicated?

(3) Describe the whole process of the addition polymerization initiated by free radicals.

(4) How are the free radicals for the initiation of addition polymerization generated? Give at least two common examples.

(5) Translate the text into Chinese.

Reading Material

Chain Reaction Polymerization

Many substances of the general formula M are able to behave as difunctional monomers by the opening of multiple bonds or strained rings and, ignoring end-groups, the resulting polymers may be described by the general formula $(M)_n$ ①.

These polymers have a repeat unit which is identical in composition to the monomer and they are formed without the loss of any portion of the monomer molecules. Polymers of this type are commonly termed addition polymers and the reactions which form them are referred to as addition or chain-growth polymerizations.

Addition polymerization invariably proceeds by a chain reaction mechanism, chain initiation being achieved by addition of an active initiator (the term catalyst is frequently but erroneously used) which reacts with the monomers to produce an active center. Addition of further monomer molecules to the resulting active centers proceeds in a series of rapid propagation steps, unit termination occurs, either by a chemical reaction of the active center or by exhaustion of the monomer supply②. The polymerization may thus be formally represented in the following way, where * denotes the active center.

$$I^* + M \rightarrow IM^* \quad \text{Initiation}$$

$$IM^* + M \rightarrow IMM^*$$

or in general

$I\text{-}(M)_n^* + M \rightarrow I-(M)_n-M^*$　　　Propagation

$I\text{-}(M)_n^* + ?$ →Inactive polymer　　　Termination

Within the limits imposed by the dependence of reactivity on monomer structure and the possibility of side reactions between initiator and monomer③, the propagating centres in addition polymerization may be free radicals, anions, cations or a variety of complex co-ordination compounds; the mechanism by which chains are terminated depends upon the nature of the active centers.

We have seen that addition polymerizations may involve a variety of active centers, and although each may lead to rather different polymerization behaviour, all such polymerizations have a number of common features. The production of polymer in each case involves the rapid addition of monomer to a few active centers, the monomer concentration decreasing slowly throughout the reaction. High molecular weight polymer is present even at low conversions, the mixture always consisting of high polymer and unreacted monomer. As with all chain reactions④, successful addition polymerization requires highly purified materials to avoid adventitious termination by impurities.

1. New words

initiation[iˌniʃi'eiʃən]*n.* 开始,引发
proceed[prɔ'sid]*vi.* 继续进行
propagation[ˌprɔpə'geiʃən]*n.* 增长,繁殖
termination[təːmi'neiʃən]*n.* 终止,端基
exhaustion[ig'zɔːstʃən]*n.* 消耗,用完
denote[di'nəut]*n.* 表示,代表
anion['ænaiən]*n.* 阴离子,负离子
cation['kætaiən]*n.* 阳离子,正离子
conversion[kən'vəːʃən]*n.* 转化(率)

2. Phrases and expressions

strained ring　张力环
chain-growth　链增长
active initiator　活性引发剂
chain initiation　链引发
co-ordination compound　络合配位化合物

3. Notes to the text

① Many substances of the general formula M are able to behave as difunctional monomers by the opening of multiple bonds or strained rings and, ignoring end-groups, the resulting polymers may be described by the general formula$(M)_n$.　通式为M的很多物质都能通过打开重键或张力环和忽略端基起到二官能团单体的作用,生成的聚合物可用通式$(M)_n$来描述。

② Addition of further monomer molecules to the resulting active centers proceeds in a series of rapid propagation steps, until termination occurs, either by a chemical reaction of the active center or by exhaustion of the monomer supply.　单体分子进一步向生

成的活性中心加成,经一系列快速增长步骤,直到发生终止,这种反应终止或是通过活性中心的化学反应,或是单体耗尽而发生。

③ Within the limits imposed by the dependence of reactivity on monomer structure and the possibility of side reactions between initiator and monomer,… 在活性对于单体结构的依赖以及引发剂和单体之间发生副反应的可能性所限定的范围之内,……

④ As with all chain reactions,… 正如所有的链反应那样,……

Lesson 13 Emulsion Polymerization(1)

The preceding discussion of free-radical addition polymerization has considered only homogeneous reactions. Considerable polymer is produced commercially by a complex heterogeneous free-radical addition process known as emulsion polymerization. This process was developed in the United States during World War II to manufacture synthetic rubber. A rational explanation of the mechanism of emulsion polymerization was proposed by Harkins and quantified by Smith and Ewart after the war, when information gathered at various locations could be freely exchanged①. Perhaps the best way to introduce the subject is to list a typical reactor charge:

Typical Emulsion Polymerization Charge

100 parts(by weight)monomer(water insoluble)

180 parts water

2 ~ 5 parts fatty acid soap

0.1 ~ 0.5 part water-soluble initiator

0 ~ 1 part chain-transfer agent(monomer soluble)

The first question is, what's the soap for? Soaps are the sodium or potassium salts of organic acids or sulfates:

$$[R-\overset{\overset{\displaystyle O}{\|}}{C}-O]^- Na^+$$

When they are added to water in low concentrations, they ionize and float around freely much as sodium chloride ions would. The anions, however, consist of a highly polar hydrophilic(water-seeking) "head" (COO]$^-$) and an organic, hydrophobic (water-fearing) "tail" (R). As the soap concentration is increased. A value is suddenly reached where the anions begin to agglomerate in micelles rather than float around individually. These micelles have dimensions on the order of 5-6 nm, far too small to be seen with a light microscope. They consist of a tangle of the hydrophobic tails in the interior(getting as far away from the water as possible) with the hydrophilic heads on the outside. This process is easily observed by following the variation of a number of solution properties with soap concentration, for example, electrical conductivity or surface tension. The break occurs when micelles start to form, and is known as the *critical micelle concentration* or CMC.

When an organic monomer is added to an aqueous micelle solution, it naturally prefers the organic environment within the micelles. Some of it congregates there, swelling the micelles until an equilibrium is reached with the contraction force of surface tension. Most of the monomer, however, is distributed in the form of much larger (1μm or 1000 nm) droplets stabilized by soap. This complex mixture is an emulsion. The cleaning action of soaps depends on their ability to emulsify oils and greases.

Despite the fact that most of the monomer is present in the droplets, the swollen micelles, because of their much smaller size, present a much larger surface area than the droplets②. This is easily seen by assuming a micelle volume to drop volume ratio of 1/10 and using the ballpark figures given above. Since the surface/volume ratio of a sphere is $3/R$.

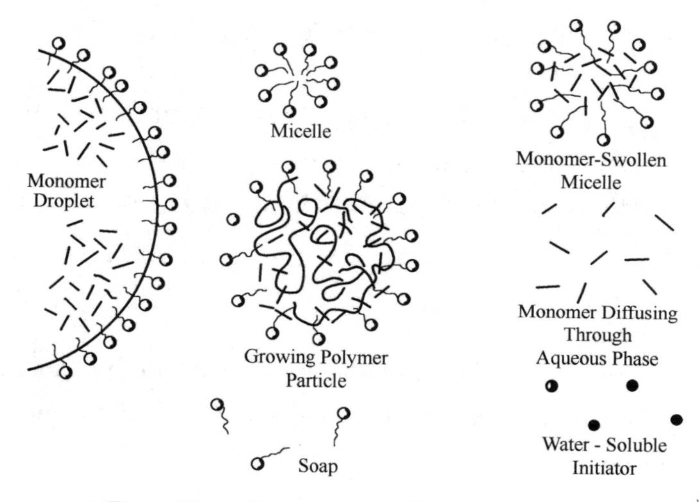

Figure 13. 1 Structures in emulsion polymerization

$$\frac{S_{\text{micelle}}}{S_{\text{droplet}}}=\left(\frac{V_{\text{mic.}}}{V_{\text{drop.}}}\right)\left(\frac{R_{\text{drop.}}}{R_{\text{mic.}}}\right)\approx\left(\frac{1}{10}\right)\left(\frac{1000}{5}\right)=20$$

Figure 13. 1 illustrates the structures present during emulsion polymerization. Free radicals for classical emulsion polymerizations are generated in the aqueous phase by the decomposition of water-soluble initiators, usually potassium or ammonium persulfate:

$$\underset{\text{Persulfate}}{S_2O_8^{2-}} \longrightarrow \underset{\text{Sulfate ion radical}}{2\,^-SO_4\cdot}$$

1. New words

emulsion[i'mʌlʃən] *n.* 乳状液

sodium['səudjəm] *n.* 钠

potassium[pə'tæsiəm] *n.* 钾

sulfate['sʌlfeit] *n.* 硫酸盐;*vt.* 以硫酸或硫酸盐处理,使变为硫酸盐

ionize['aiʌnaiz] *vt.* 使离子化;*vi.* 电离

hydrophilic[ˌhaidrəu'filik] *adj.* 亲水的,吸水的

agglomerate[ə'glɔməreit] *n.* 大团,大块

micelle[mi'sel] *n.* 胶束,胶囊,微胞,微团,胶态离子

sphere[sfiə] *n.* 球,球体,范围,领域,方面,圈子,半球

2. Phrases and expressions

fatty acid　脂肪酸
synthetic rubber　合成橡胶
aqueous phase　液相
ammonium persulfate　过(二)硫酸铵

3. Notes to the text

① A rational explanation of the mechanism of emulsion polymerization was proposed by Harkins and quantified by Smith and Ewart after the war, when information gathered at various locations could be freely exchanged.　战后 Harkins 为乳液聚合的机理提出了一种合理的解释,并由 Smith 和 Ewart 加以量化,因为此时在不同地方收集到的信息能够自由交换。

② Despite the fact that most of the monomer is present in the droplets, the swollen micelles, because of their much smaller size, present a much larger surface area than the droplets.　尽管实际中的单体大多都是呈现小滴状的,但是膨胀微胞比小滴具有更大的表面积,因为它们的尺寸更小。

4. Exercises

(1) What was the reason for the development of emulsion polymerization?

(2) Why could the rational explanation of the mechanism of emulsion polymerization be proposed and quantified?

(3) What's the soap for in the typical emulsion polymerization?

(4) What's the purpose of following the variation of a number of solution properties with soap concentration?

(5) What does the CMC mean?

Reading Material

Emulsion Polymerization (2)

Most emulsion polymerizations are initiated by free-radicals and show many of the characteristics of other free-radical systems, the main mechanistic differences resulting from the localization of the propagating free-radicals in small volume elements. Polymerization in emulsion not only permits the synthesis of polymers of high molecular weight at high rates, often in cases where bulk polymerization is not possible or is inefficient, but has other important practical advantages. The water absorbs much of the heat of polymerization and facilitates reaction control, and the product is obtained in a relatively non-viscous and easily handled form, to be used as such or readily to be converted into solid polymer after coagulation, washing and drying①. In copolymerization, although copolymerization theory is applicable to emulsion systems, differences in solubility of the monomers in the aqueous phase may result in deviations from the behaviour in bulk, with important practical consequences.

Emulsion polymerizations are open to great variation②, from simple systems containing a single monomer, emulsifier, water and a simple initiator to those containing two or three monomers, added at once or in increments, mixed emulsifiers and auxiliary stabilizers, and complex initiator systems including chain transfer agents.

The ratio of monomer to aqueous phase can be varied widely but in technical practice it is generally limited to the range 30/70 to 60/40. The higher levels approach the limit which can be achieved by direct polymerization and owing to the substantial amount of heat to be dissipated can often only be obtained by incremental addition of monomer③. A further complication is that the latex increases substantially in viscosity, particularly with the more water-soluble monomers which are miscible with the polymer, and then falls as the reaction is taken to completion. This stage is not infrequently accompanied by an increase in the latex particle size by aggregation and, when surface stabilization is inadequate, by coagulation④.

1. New words

facilitate[fə'siliteit] *vt.* 使容易,便于
coagulation[kəuˌægju'leiʃən] *n.* 凝结
copolymerization[kəuˌpɔliməri'zeiʃən] *n.* 共聚
aqueous['eikwiəs] *adj.* 含水的,水的
deviation[ˌdi:vi'eiʃən] *n.* 偏差,偏离
emulsifier[i'mʌlsifaiə] *n.* 乳化剂
auxiliary[ɔ:g'ziljəri] *adj.* 辅助的,副的
stabilizer['steibi'laizə] *n.* 稳定剂
substantial[səb'stænʃəl] *adj.* 很多的,物质的
dissipate['disipeit] *vt.* 分散,消散
latex['leiteks] *n.* 胶乳,乳液
miscible['misibl] *adj.* 易混的

2. Phrases and expressions

emulsion polymerization 乳液聚合
bulk polymerization 本体聚合
copolymerization theory 共聚理论
in increment 分批量,分期地
auxiliary stabilizer 辅助稳定剂
complex initiator system 复合引发体系
chain transfer agent 链转移剂
owing to 因为,由于,归因于
water-soluble 水溶性的

3. Notes to the text

① The water absorbs much of the heat of polymerization and facilitates reaction control, and the product is obtained in a relatively non-viscous and easily handled form, to be used as such or readily to be converted into solid polymer after coagulation, washing and drying. 水能吸收大量的聚合热并有利于反应的控制,产品可以以一种相对无黏度

并易于处理的形式获得,可以直接使用也可以经凝聚、洗涤和干燥后转化为固体聚合物。

② Emulsion polymerizations are open to great variation,… 乳液聚合容易发生很大的变化,……

③ The higher levels approach the limit which can be achieved by direct polymerization and owing to the substantial amount of heat to be dissipated can often only be obtained by incremental addition of monomer.

较高的水平达到了极限,这个极限可以通过直接聚合达到,同时由于要放出大量的热量,这个极限只能用分批补加单体的办法来获得。

④ This stage is not infrequently accompanied by an increase in the latex particle size by aggregation and when surface stabilization is inadequate by coagulation. 这一阶段常常会出现由于聚集作用和由凝结造成的表面不稳定而导致的胶乳粒子增大的现象。

Lesson 14 Ionic Polymerization

Strong Lewis acids, that is, electron acceptors, are often capable of initiating addition polymerization of monomers with electron-rich substituents adjacent to the double bond①. Cationic catalysts are most commonly metal trihalides such as $AlCl_3$ or BF_3. These compounds, although electrically neutral, are two electrons short of having a complete valence shell of eight electrons. They were found to require traces of a cocatalyst, usually water, to initiate polymerization, first by grabbing a pair of electrons from the cocatalyst:

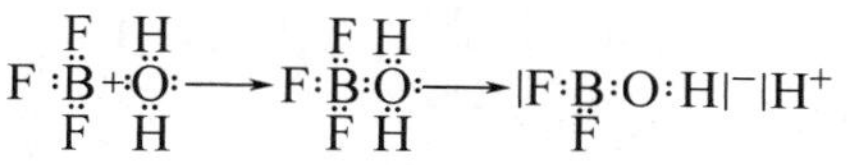

$$F{:}\overset{F}{\underset{F}{B}} + {:}\overset{H}{\underset{H}{O}}{:} \longrightarrow F{:}\overset{F}{\underset{F}{B}}{:}\overset{H}{\underset{H}{O}}{:} \longrightarrow |F{:}\overset{}{\underset{F}{B}}{:}O{:}H|^{-}|H^{+}$$

The leftover proton is thought to be the actual initiating species, abstracting a pair of electrons from the monomer and leaving a cationic chain end which reacts with additional monomer molecules.

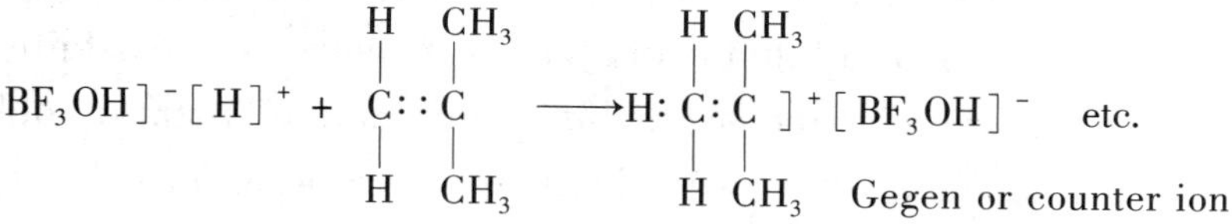

$$BF_3OH]^-[H]^+ + \overset{H}{\underset{H}{C}}{::}\overset{CH_3}{\underset{CH_3}{C}} \longrightarrow H{:}\overset{H}{\underset{H}{C}}{:}\overset{CH_3}{\underset{CH_3}{C}}\ ]^+[BF_3OH]^- \quad \text{etc.}$$

Gegen or counter ion

An important point here is that the *gegen or counter* ion is electrostatically held near the growing chain end and so can exert a steric influence on the addition of monomer units②. Termination is thought to occur by a disproportionation-like reaction which regenerates the catalyst complex. The complex, therefore, is a true catalyst, unlike free-radical initiators:

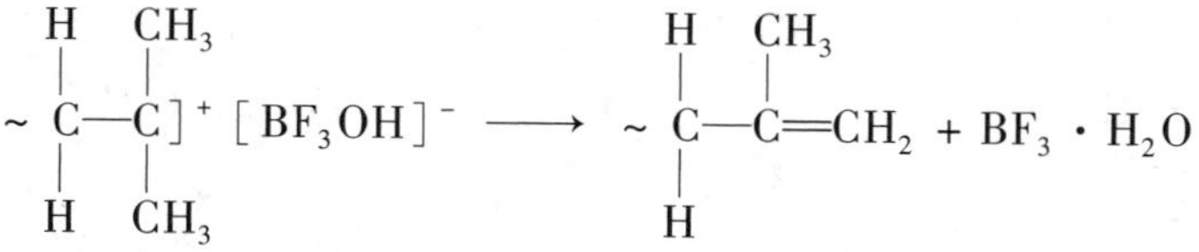

$$\sim \overset{H}{\underset{H}{C}}-\overset{CH_3}{\underset{CH_3}{C}}]^+[BF_3OH]^- \longrightarrow \sim \overset{H}{\underset{H}{C}}-\overset{CH_3}{C}{=}CH_2 + BF_3 \cdot H_2O$$

The kinetics of these reactions is not well understood, but they proceed very rapidly at low the polymerization of isobutylene illustrated above is carried out commercially at -101.5℃. The average chain length increases as the temperature is lowed.

Cationic initiation is successful only with monomers like isobutylene having electron-rich substituents adjacent to the double bond, such as

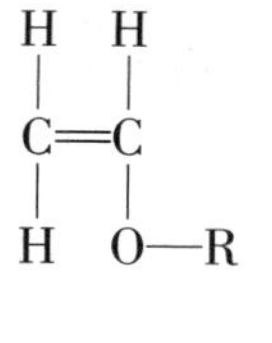

$$\overset{H}{\underset{H}{C}}{=}\overset{H}{\underset{O-R}{C}} \qquad \overset{H}{\underset{H}{C}}{=}\overset{CH_3}{\underset{C_6H_5}{C}}$$

Alkyl vinyl ethers 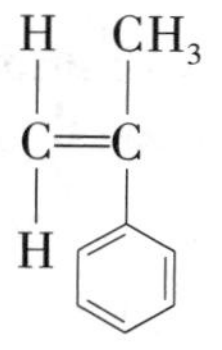 α-Methyl Styrene

None of these monomers can be polymerized to high molecular weight with free-radical initiators.

Addition polymerization may also be initiated by anions. Anionic polymerization has achieved tremendous commercial importance in the past two decades because of its ability to control molecular structure during polymerization, allowing the synthesis of materials that were previously difficult or impossible to obtain. A variety of anionic initiators has been investigated, but the organic alkali-metal salts are perhaps most common, as illustrated below for the polymerization of styrene with *n*-Butyllithium:

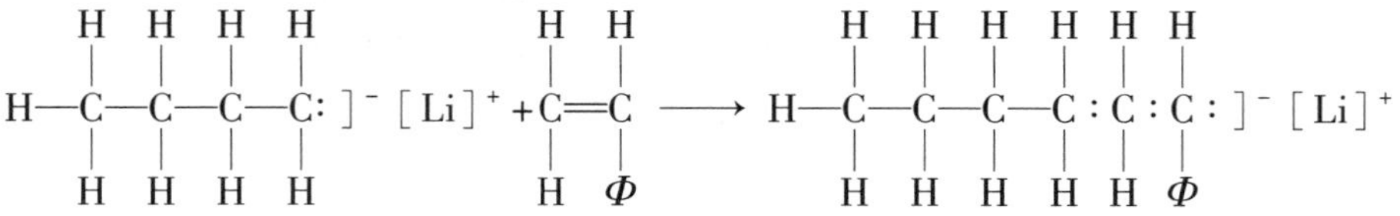

$$\begin{array}{ccccccccccccccccccccccccc}
 & H & & H & & H & & H & & H & & H & & H & & H & & H & & H & & H & & H & \\
 & | & & | & & | & & | & & | & & | & & | & & | & & | & & | & & | & & | & \\
H- & C & - & C & - & C & - & C & :]^-[Li]^+ + & C & = & C & \longrightarrow H- & C & - & C & - & C & - & C & : & C & : & C & :]^-[Li]^+ \\
 & | & & | & & | & & | & & | & & | & & | & & | & & | & & | & & | & & | & \\
 & H & & H & & H & & H & & H & & \Phi & & H & & H & & H & & H & & H & & \Phi &
\end{array}$$

n-Butyllithium (*n*-BuLi)   Styrene

The anionic chain end then propagates the chain by adding another monomer molecule. Again, the *gegen ion* can sterically influence the reaction.

Sodium and lithium *metals* were used to polymerize butadiene in Germany during World War II. After the war, in the United States, it was discovered that under appropriate conditions, dispersions of lithium could lead to largely cis-1,4 addition of butadiene and isoprene (the latter being the synthetic counterpart of natural rubber). In these processes, a metal atom first reacts with the monomer to form an anion radical:

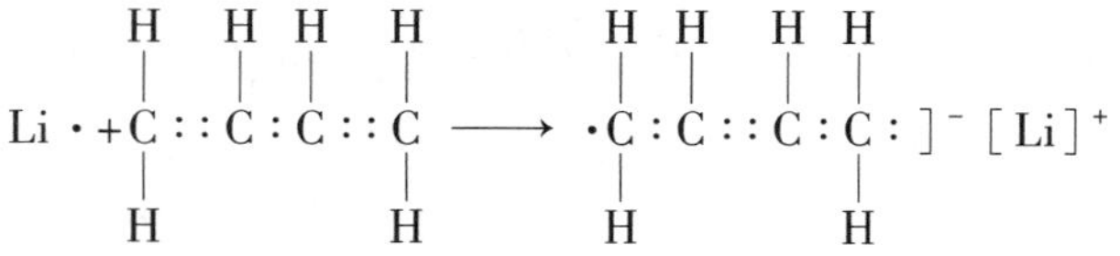

$$\begin{array}{ccccccccccccccccc}
 & H & & H & & H & & H & & H & & H & & H & & H & \\
 & | & & | & & | & & | & & | & & | & & | & & | & \\
Li\cdot + & C & :: & C & : & C & :: & C & \longrightarrow \cdot & C & : & C & :: & C & : & C & :]^-[Li]^+ \\
 & | & & & & & & | & & | & & & & & & | & \\
 & H & & & & & & H & & H & & & & & & H &
\end{array}$$

Anion radical

These anion radicals then react in either of two ways. One may react with another atom of lithium,

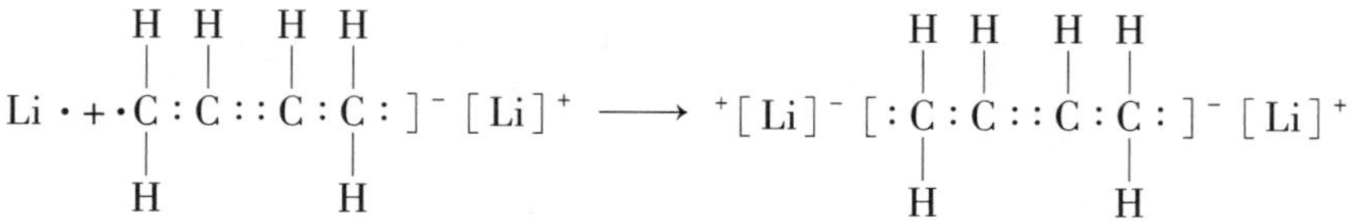

$$\begin{array}{ccccccccccccccccc}
 & H & & H & & H & & H & & H & & H & & H & & H & \\
 & | & & | & & | & & | & & | & & | & & | & & | & \\
Li\cdot + \cdot & C & : & C & :: & C & : & C & :]^-[Li]^+ \longrightarrow {}^+[Li]^-[: & C & : & C & :: & C & : & C & :]^-[Li]^+ \\
 & | & & & & & & | & & | & & & & & & | & \\
 & H & & & & & & H & & H & & & & & & H &
\end{array}$$

and/or two may rapidly undergo radical recombination,

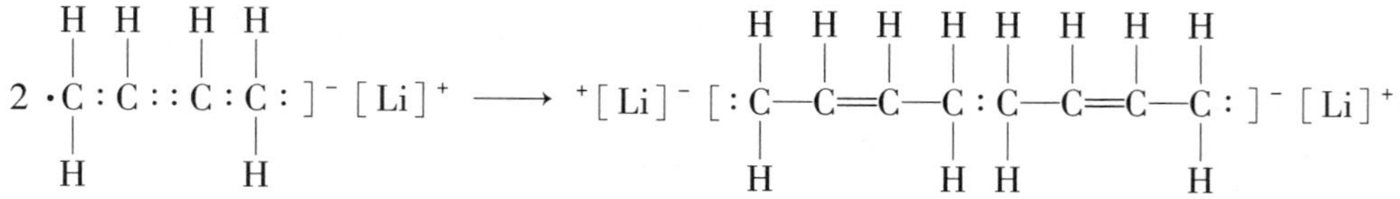

$$\begin{array}{ccccccccccccccccccccccccc}
 & H & & H & & H & & H & & H & & H & & H & & H & & H & & H & & H & & H & \\
 & | & & | & & | & & | & & | & & | & & | & & | & & | & & | & & | & & | & \\
2\cdot & C & : & C & :: & C & : & C & :]^-[Li]^+ \longrightarrow {}^+[Li]^-[: & C & - & C & = & C & - & C & : & C & - & C & = & C & - & C & :]^-[Li]^+ \\
 & | & & & & & & | & & | & & & & & & | & & | & & & & & & | & \\
 & H & & & & & & H & & H & & & & & & H & & H & & & & & & H &
\end{array}$$

Either way, the result is a dianion that propagates a chain from each end. Other dianionic initiators have been developed.

1. New words

catalyst[ˈkætəlist] *n.* 催化剂

trihalides[ˌtraiˈhælaid] *n.* 三卤化合物

neutral['nju:trəl] *adj.* 中性的,无确定性质的,(颜色等)不确定的
cocatalyst[kəu'kætəlist] *n.* 助催化剂
electrostatically [i'lektrəu'stætiks] *n.* 静电学
isobutylene[ˌaisəu'bju:tili:n] *n.* [化]异丁烯
dianion[dai'ænaiən] *n.* 二价阴离子
butyllithium['bju:til'liθiəm] *n.* 丁基锂

2. Phrases and expressions

leftover proton 过剩的质子
catalyst complex 催化剂混合物
gegen or counter ion 反离子,抗衡离子
polar solvent 极性溶剂

3. Notes to the text

① Strong Lewis acids, that is, electron acceptors, are often capable of initiating addition polymerization of monomers with electron-rich substituents adjacent to the double bond. 作为电子接受体,强路易斯酸时常可以致使双键附近有富含电子取代基的聚合物发生加成聚合反应。

② An important point here is that the gegen or counter ion is electrostatically held near the growing chain end and so can exert a steric influence on the addition of monomer units. 反离子是靠静电力而得以存在于生长分子链的链端的,并因此能对单体单元的加成反应施加阻滞性影响,这一点很重要。

4. Exercises

(1) Translate the text into Chinese.

(2) What're the characters of Cationic catalysts?

(3) What's the reason for anionic polymerization to achieve tremendous commercial importance in the past two decades?

(4) Give a description of the process of anionic polymerization.

Reading Material

Bulk Polymerization

The simplest and most direct method of converting monomer to polymer is known as *bulk* or mass polymerization. A typical charge for a free-radical bulk polymerization might consist of a liquid monomer, a monomer-soluble initiator, and perhaps a chain-transfer agent.

As simple as this seems, some serious difficulties can be encountered, particularly in free radical bulk polymerizations. One of them is illustrated in Figure 14. 1, which indicates the course of polymerization for various concentrations of methyl methacrylate in benzene, an inert solvent. The reactions were carefully maintained at constant temperature. As polymer concentrations increase, however, a distinct acceleration of the rate of polymerization is observed which does not conform to the classical kinetic scheme. This phenomenon is known variously as *autoacceleration*, the *gel effect*, or the *Tromsdorff effect*.

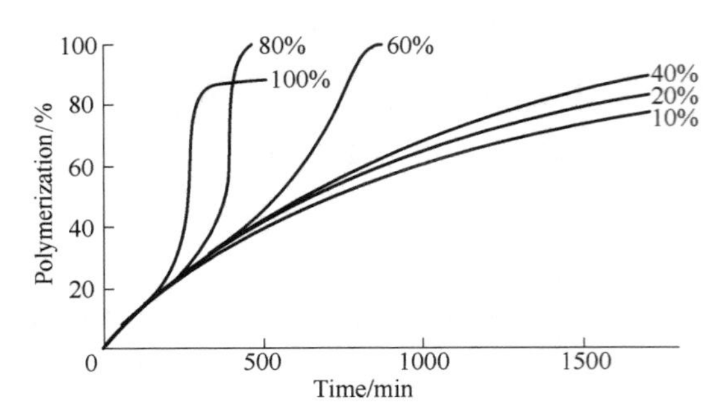

Figure 14. 1 Polymerization of methyl methacrylate at 50℃ in the presence of benzoyl peroxide at various concentrations of monomer in benzene

The reasons for this behavior lie in the difference between the propagation reaction and the termination steps, and the extremely high viscosity of concentrated polymer solutions (10^4 poise might be a ballpark figure). The propagation reaction involves the approach of a small monomer molecular to a growing chain end, whereas termination requires that the ends of two growing chains get together. At high concentrations of polymer, it becomes exceedingly difficult for the growing chain ends to drag their chains through the entangled mass of dead polymer chains①. It is nowhere near as difficult for a monomer molecule to pass through the reaction mass. Thus, the rate of the termination reaction is limited not by the nature of the chemical reaction, but by the rate at which the reactants can diffuse together; that is, it is *diffusion controlled*②. This lowers the effective termination rate constant k_t, and since k_t appears in the denominator of, the net effect is to increase the rate of polymerization. At very high polymer concentrations and below the temperature at which the chains become essentially immobile-Tg of the monomer plasticized polymer-even the propagation reaction is diffusion limited, hence the leveling off of the 100% curve.

The difficulties are compounded by inherent nature of the reaction mass.

1. New words

autoacceleration[ækˌsel'əreiʃən] *n.* 自动加速效应

methacrylate[me'θækrəleit] *n.* 甲基丙烯酸酯

viscosity[vis'kɔseti] *n.* 黏度,黏性

benzoyl[benzəuil] *n.* 苯甲酰基

peroxide[per'ɔksaid] *n.* 过氧化物

2. Phrases and expressions

chain-transfer agent 链转移剂

methyl methacrylate in benzene 甲基丙烯酸苯

inert solvent 非极性溶剂

the gel effect 凝胶效应

propagation reaction 链增长反应

termination reaction 终止反应

benzoyl peroxide 过氧化苯甲酰

3. Notes to the text

① At high concentrations of polymer, it becomes exceedingly difficult for the growing chain ends to drag their chains through the entangled mass of dead polymer chains.
在高浓度聚合物中,增长中的分子链将它们的链段从大量互相缠结的已失活的聚合物分子链中拖拽出来是极端困难的。

② Thus, the rate of the termination reaction is limited not by the nature of the chemical reaction, but by the rate at which the reactants can diffuse together; that is, it is diffusion controlled. 因此,链终止反应的速率不仅受到化学反应类型的限制,还受到反应物相互扩散能力的限制;也就是说,链终止反应的速率是受扩散控制。

PART 4 POLYMER PROPERTIES

Lesson 15 Rubber Elasticity

Natural and synthetic rubbers possess some interesting, unique, and useful mechanical properties. No other materials are capable of reversible extension of 600% ~ 700%. No other materials exhibit an increase in modulus with increasing temperature. It was recognized long ago that vulcanization was necessary for rubber deformation to be completely reversible. We now know that this is a result of the crosslinks so introduced preventing the bulk slippage of the molecules past one another, eliminating flow (irrecoverable deformation)[1]. More recently, this function of the covalent crosslinks has been assumed by rigid domains (either glassy or crystalline) within some linear polymers. Thus, when a stress is applied to a sample of crosslinked rubber, equilibrium is established fairly rapidly. Once at equilibrium, the properties of the rubber can be described by thermodynamics[2].

Thermodynamics of Elasticity

Consider an element of material with dimensions $a \times b \times c$, as sketched in Figure 15.1. Applying the first law of thermodynamics to this system yields

$$dU = dQ - dW \qquad (15.1)$$

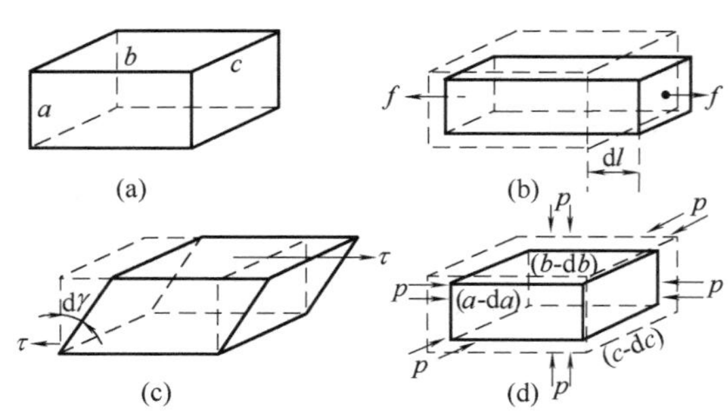

Figure 15.1 Types of mechanical deformation: (a) unstressed; (b) uniaxial tension; (c) pure shear; (d) isotropic compression

Where dU is the change in the system's internal energy, and dQ and dW are the heat and work exchanged between system and surroundings as the system undergoes a differential change. (We have adopted the convention here that work done by the system on the surroundings is positive.)

We will consider three types of mechanical work:

1. Work done by a uniaxial tensile force f:

$$dW(\text{tensile}) = -f dl \qquad (15.2)$$

Where dl is the differential change in the system's length arising from the application of the force f. This is the fundamental definition of work. The negative sign arises from the need to reconcile the mechanical convention of treating a tensile force (which does work on the system) as positive with the thermodynamic convention above.

2. Work done by a shear stress τ:

$$dW(\text{shear}) = (\text{force})(\text{distance}) = -(\tau bc)(a d\gamma) = -\tau V d\gamma \tag{15.3}$$

Where γ is the shear strain (Figure 12.1 c) and $V = abc =$ the system volume.

3. Work done by an isotropic pressure in changing the volume:

$$dW(\text{pressure}) = p(cb)da + p(ac)db + p(ab)dc = pdV \tag{15.4}$$

Note that no minus sign is needed here. A positive pressure causes a decrease in volume (negative dV) and does work on the system.

If the deformation process is assumed to occur reversibly (in a thermodynamic sense), then

$$dQ = TdS \tag{15.5}$$

Where S is the system's entropy.

Combining the preceding five equations gives a general relation for the change of internal energy of an element of material undergoing a differential deformation:

$$dU = TdS - pdy + fdl + V\tau d\gamma \tag{15.6}$$

1. New words

unique [juː'niːk] *adj.* 独一无二的，特有的

vulcanization [ˌvʌlkənai'zeiʃən] *n.* (橡胶的)硫化

reversible [ri'vəːsəbl] *adj.* 可逆的

covalent [kəu'veilənt] *adj.* 共有原子价的，共价的

equilibrium [ˌiːkwi'libriəm] *n.* 平衡，均势

uniaxial ['juːni'æksiəl] *adj.* 单轴的，单轴晶体

thermodynamics [ˌθəːməudai'næmiks] *n.* 热力学

dimension [di'menʃən] *n.* 尺寸，度量，方面，部分

2. Phrases and expressions

synthesize rubber 合成橡胶

covalent crosslink 共价键交联

uniaxial tensile 单轴拉伸

3. Notes to the text

① We now know that this is a result of the crosslinks so introduced preventing the bulk slippage of the molecules past one another, eliminating flow (irrecoverable deformation). 我们现在知道这是由于交联阻止了分子链彼此间的相对滑移和流动(即不可逆形变)。

② Thus, when a stress is applied to a sample of crosslinked rubber, equilibrium is established fairly rapidly. Once at equilibrium, the properties of the rubber can be described by thermodynamics. 因此，当将外力施加在已交联的橡胶试样上时，平衡态便迅速

建立起来。一旦处于平衡态时,橡胶的性质便可以用热力学来描述。

4. Exercises

(1) Describe some special properties of natural and synthetic rubber.

(2) This chapter has emphasized the differences between the behavior of polymers and nonpolymers in uniaxial tension. How would you expect them to compare in isotropic compression? Why?

(3) What does it yield when you apply the first law of thermodynamics to an element of material with a 3-dimension system?

(4) When a sample of material is affected by stress, which types of mechanical work should be considered.

Reading Material

Purely Viscous Flow

Rheology is the science of the deformation and flow of materials. Much contemporary rheological work is concerned directly with polymer systems because they exhibit such interesting, unusual, and difficult-to-describe (at least from the standpoint of traditional materials) deformation behavior. The simple and traditional linear engineering models, Newton's law (for flow) and Hooke's law (for elasticity), often just aren't reasonable approximations. Not only are the elastic and viscous properties of polymer melts and solutions usually nonlinear, but they exhibit a combination of viscous and elastic response, the relative magnitudes of which depend on the temperature and the time scale of the experiment①. This *viscoelastic* response is dramatically illustrated by Silly Putty (a silicone polymer). When bounced (stressed rapidly), it is highly elastic, recovering most of the potential energy it had before being dropped. If stuck on the wall (stressed over a long time period), however, it will slowly flow down the wall, albeit with a high viscosity, and will show little tendency to recovery any deformation.

We limit ourselves here to two-dimensional deformations. A detailed three-dimensional treatment of rheology is beyond the scope of this book. Several excellent treatises are available.

We begin our treatment of rheology with a discussion of purely viscous flow. For our purposes, this will be defined as a deformation process in which all the applied mechanical energy is nonrecoverably dissipated as heat in the material through molecular friction. This process is known as viscous energy dissipation. Purely viscous flow is in most cases a good approximation for dilute polymer solutions, and often for concentrated solutions and melts where the stress on the material is not changing too rapidly; that is, where equilibrium flow is approached②.

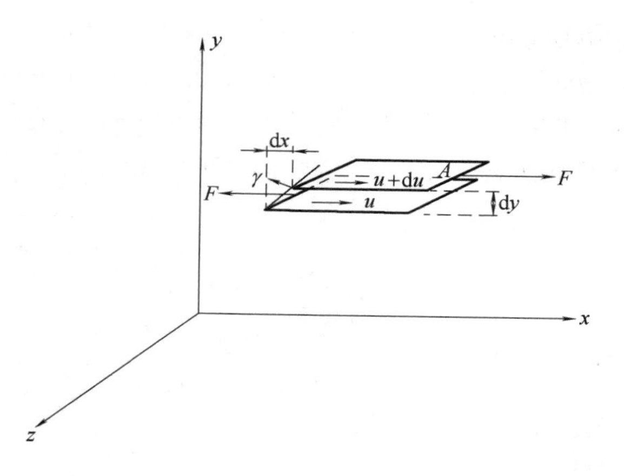

Figure 15.2 Definitions of shear stress and shear rate

The *viscosity* of a material expresses its resistance to flow. It is defined quantitatively in terms of two basic parameters; the *shear stress* τ and the *shear rate* (more correctly, the rate of shear straining) $\dot{\gamma}$. These quantities are defined in Figure 15.2. Consider a point in a laminar flow field (Figure 15.2). A rectangular coordinate system is established with the x axis (sometimes designated the 1 coordinate direction) in the direction of flow, and the y axis (the 2 direction) perpendicular to surfaces of constant fluid velocity, that is, parallel to the *velocity gradient*. The z axis (3 or neutral direction) is mutually perpendicular to the others. This is known as *simple shearing* flow. It is one example of a *viscometric flow*, a flow field in which the velocity and its gradient are everywhere perpendicular, with a third neutral direction mutually perpendicular to the others. Viscometric flows can often be treated analytically. Fortunately, many laminar flow situations encountered in practice either are viscometric flows or least can reasonably be approximated by them.

1. New words

rheological[ˌriə'lɔdʒikl] *adj.* 流变学的
contemporary[kən'tempərəri] *adj.* 当代的,同时代的
viscoelastic[ˌviskəui'læstik] *adj.* 黏弹性的
albeit[ɔːl'biːt] *conj.* 尽管,即使
rigorous['rigərəs] *adj.* 严密的,缜密的
laminar['læminə] *adj.* 薄片状的,层状的

2. Phrases and expressions

velocity gradient 速度梯度
viscometric flow 黏性流动
laminar flow 层流
simple shearing flow 简单剪切流

3. Notes to the text

① Not only are the elastic and viscous properties of polymer melts and solutions usually nonlinear, but they exhibit a combination of viscous and elastic response, the relative magnitudes of which depend on the temperature and the time scale of the experiment. 通常聚合物融体和溶液的黏性和弹性不仅是非线性关系的,并且它们表现出二者相

结合的一种反应,这种结合的相对程度取决于实验温度和时间。

② Purely viscous flow is in most cases a good approximation for dilute polymer solutions, and often for concentrated solutions and melts where the stress on the material is not changing too rapidly; that is, where equilibrium flow is approached. 在多数情况下,纯黏性流动与聚合物的稀溶液非常相似,而对于浓溶液和聚合物熔体来说,这些材料在外力作用下的变化不很迅速,接近于平衡流动。

Lesson 16 Flow Curves of Polymer Fluid

When most materials are subjected to a constant shear rate (or constant shear stress) at a fixed temperature, a corresponding steady-state value of shear stress (or shear rate) is soon established①. The steady-state relation between shear stress and shear rate at constant temperature is known as a *flow curve*.

Newton's "law" of viscosity states that the shear stress is linearly proportional to the shear rate, the proportionality constant being the viscosity η:

$$\tau = \eta\dot{\gamma} \tag{16.1}$$

Fluids that obey this hypothesis are termed Newtonian. The hypothesis holds quite well for many nonpolymer fluids, such as gases, water, and toluene. This type of flow behavior would be expected for small, relatively symmetrical molecules, where the structure and/or orientation do not change with the intensity of shearing②.

An arithmetic flow curve (τ vs. $\dot{\gamma}$) for a Newtonian fluid is a straight line through the origin with a slope η [Figure 16.1(a)]. Because τ and $\dot{\gamma}$ often cover very wide ranges, it is usually preferable to plot them on log-log coordinates. Taking logarithms of both sides of Equation 16.1 yields

$$\log\tau = \log\eta + \log\dot{\gamma} \tag{16.2}$$

Hence, a log-log plot of τ vs. $\dot{\gamma}$, a logarithmic flow curve, will be a line of slope unity for a Newtonian fluid [Figure 16.1(b)].

Unfortunately, many fluids do not obey Newton's hypothesis. Both *dilatants* (shear-thickening) and pseudoplastic (shear-thinning) fluids have been observed (Figure 16.1). On log-log coordinates, dilatants flow curves have a slope greater than 1 and

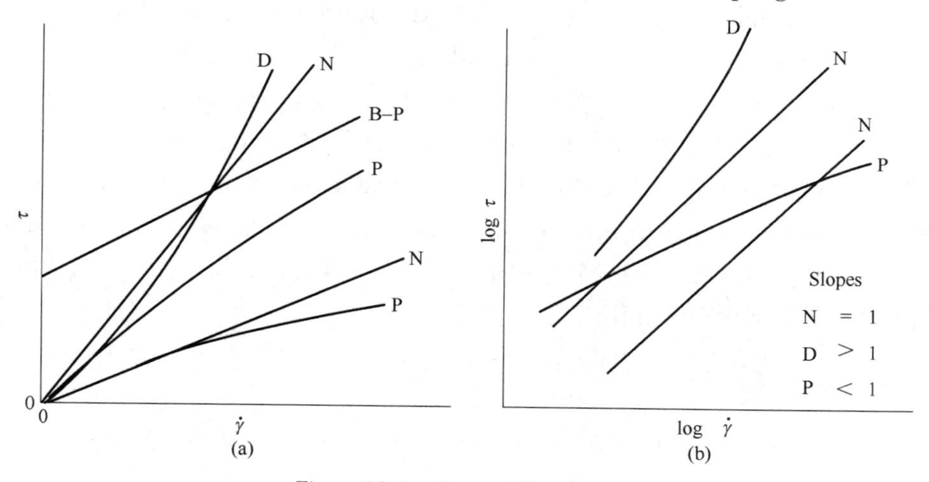

Figure 16.1 Types of flow curves

(a) arithmetic (b) logarithmic N—Newtonian P—pseudoplastic B-P—Bingham plastic (infinitely pseudoplastic) D—dilatant

pseudoplastics have a slope less than 1. Dilatant behavior is reported for certain slurries and implies an increased resistance to flow with intensified shearing. Polymer melts and solutions are invariably pseudoplastic. That is, their resistance to flow decreases with the intensity of shearing.

For non-Newtonian fluids, since τ is not directly proportional to $\dot{\gamma}$, the viscosity is not constant. Plots (or equations) giving η as a function of $\dot{\gamma}$ (or τ) are an equivalent method of representing a material's equilibrium viscous shearing properties. A knowledge of the relation between any two of the three variables (τ, η and $\dot{\gamma}$) completely defines the equilibrium viscous shearing behavior, since they have some relationship.

1. New words

proportional [prə'pɔ:ʃənl] *n.* 比例的，成比例的
toluene ['tɔljui:n] *n.* 甲苯
arithmetic [ə'riθmətik] *n.* 算术，算法
logarithmic [ˌlɔgə'riθmik] *adj.* 对数的
pseudoplastic [psju:dəu'plæstik] *adj.* 假塑性的
dilatant [dai'leitənt] *n.* 膨胀物 *adj.* 膨胀的，因膨胀而变形的
slurries ['slə:ri] *n.* 泥浆
symmetrical [si'metrikəl] *adj.* 对称的，匀称的
orientation [ˌɔ:rien'teiʃən] *adj.* 方向，目标

2. Phrases and expressions

steady-state　不变的状态，平衡状态，稳定状态
shear-thickening　剪切增稠
shear-thinning　剪切变稀
Bingham plastic　宾汉塑性

3. Notes to the text

① When most materials are subjected to a constant shear rate (or constant shear stress) at a fixed temperature, a corresponding steady-state value of shear stress (or shear rate) is soon established.　当多数材料在固定的温度下具有恒定的剪切速率（或恒定的剪切应力）时，一个相应的稳态剪切应力值（或剪切速率）被立即建立起来。

② This type of flow behavior would be expected for small, relatively symmetrical molecules, where the structure and/or orientation do not change with the intensity of shearing.　这种类型流体的流动行为适用于那些小的、相对对称的分子，它们的结构和取向度并不随着剪切的增强而产生变化。

4. Exercises

(1) How do you use Newton's law to express the relationship among the shear stress, shear rate and viscosity?

(2) Give some examples of polymers which are suitable for Newton's law.

(3) Describe the difference between Newtonian and non-Newtonian fluids with the curves in Figure 16. 1.

(4) On log-log coordinates, dilatants flow curves have a slope ________.

a. greater than 1 b. equal to 1

c. less than 1

(5) In non-Newtonian fluids, the viscosity is ________.

a. constant b. not constant

Reading Material

Time-Dependent Fluid Behavior

The types of non-Newtonian flow just described, though shear dependent, are time independent. As long as a constant shear rate is maintained, the same shear stress or viscosity will be observed at equilibrium. Some fluids exhibit *reversible* time—dependent properties, however. When sheared at a constant rate or stress, the viscosity of a *thixotropic* fluid will decrease over a period of time (Figure 16. 2), implying a progressive breakdown of structure. If the shearing is stopped for a while, the structure reforms, and the experiment may be duplicated. The ketchup that splashes all over after a period of vigorous tapping is a classic example. Thixotropic behavior is important in the paint industry, where smooth, even application with brush or roller is required, but it is desirable for the paint on the surface to "set up" to avoid drips and runs after application①. The opposite sort of behavior is manifested by *rheopectic* fluids, for example, certain drilling muds used by the petroleum industry. When subjected to continuously increasing and then decreasing shearing, time-dependent fluids give flow curves as in Figure 16. 3.

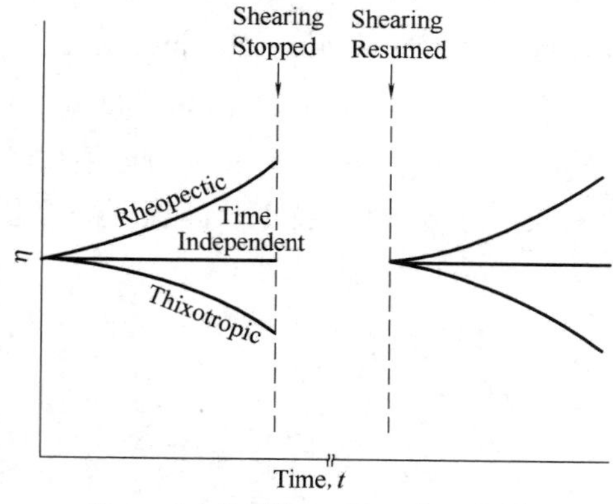

Figure 16. 2 Time-dependent fluids

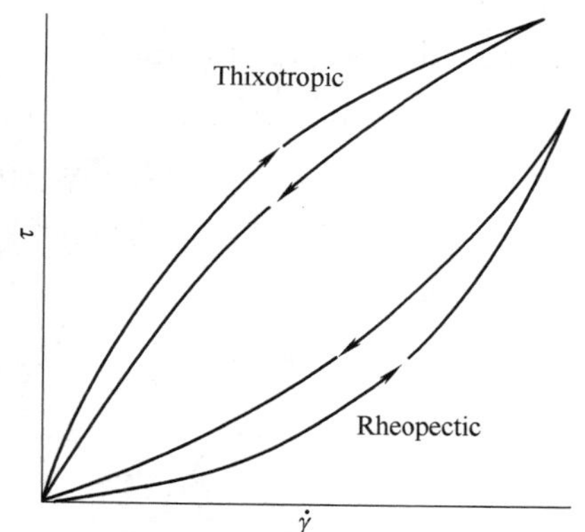

Figure 16. 3 Flow curves for time-dependent fluids under continuously increasing and then decreasing shear

Once in a while, polymer systems will appear to be thixotropic or rheopectic. Careful checking (including before and after molecular weight determinations) invariably shows that the phenomenon is not reversible and is due to degradation or crosslinking of the polymer when in the viscometer for long periods of time, particularly at elevated temperatures②. Other transient time-dependent effects in polymers are due to elasticity, and will be considered later, but for chemically stable polymer melts or solutions, the equilibrium viscous properties are time independent. We treat only such systems from here on.

1. New words

thixotropic [ˌθiksə'trɔpik] *adj.* 触变性,摇溶性
rheopectic [ˌriːə'pektik] *adj.* 振凝的,抗流变的
ketchup ['ketʃəp] *n.* 番茄酱
splash [splæʃ] *n.* 溅,飞溅,斑点;*v.* 溅,泼,溅湿
roller ['rəulə] *n.* 辊子,滚轴
vigorous ['vigərəs] *adj.* 有力的,精力充沛的
invariably [in'vɛəriəbli] *adv.* 不变地,总是
viscometer [vis'kɔmitə] *n.* 黏度计

2. Phrases and expressions

time-dependent 有时间依赖性
drilling muds 钻井泥浆

3. Notes to the text

① Thixotropic behavior is important in the paint industry, where smooth, even application with brush or roller is required, but it is desirable for the paint on the surface to "set up" to avoid drips and runs after application. 触变行为对油漆工业非常重要,油漆要使用刷子和辊子平滑均匀地涂刷,但是,表面的油漆最好能"定型"以防止流挂。

② Careful checking (including before and after molecular weight determinations) invariably shows that the phenomenon is not reversible and is due to degradation or crosslinking of the polymer when in the viscometer for long periods of time, particularly at elevated temperatures. (在相对分子质量测定前后)仔细检查的结果都表明这种现象是不可逆的,而且出现这种现象的原因是由于聚合物在黏度计中长时间、特别是高温下停留会发生降解或交联。

Lesson 17 Polymer Melts and Solutions

When the flow properties of polymer melts and solutions can be measured over a wide enough range of shearing, the logarithmic flow curves appear as in Figure 17. 1. It is generally observed that:

1. At low shear rates (or stresses), a "lower Newtonian" region is reached with a so-called *zero-shear viscosity* η_0.

2. Over several decades of intermediate shear rates, the material is pseudoplastic.

3. At very high shear rates, an "upper Newtonian" region, with viscosity η_∞ is attained.

This behavior can be rationalized in terms of molecular structure. At low shear, the randomizing effect of the thermal motion of the chain segments overcomes any tendency toward molecular alignment in the shear field①. The molecules are thus in their most random and highly entangled state, and have their greatest resistance to slippage (flow). As the shear is increased, the molecules will begin to untangle and align in the shear field, reducing their resistance to slippage past one another. Under severe shearing, they will be pretty much completely untangled and aligned, and reach a state of minimum resistance to flow. This is illustrated schematically in Figure 17. 1.

Intense shearing eventually leads to extensive breakage of main-chain bonds, that is, mechanical degradation②. Furthermore, differentiation of another situation with respect to time reveals that the rate of viscous energy dissipation per unit volume is equal to $\tau\dot{\gamma}$. It thus becomes exceedingly difficult to maintain the temperature constant under intense shearing, so good data that illustrate the upper-Newtonian region are relatively rare, particularly for polymer melts.

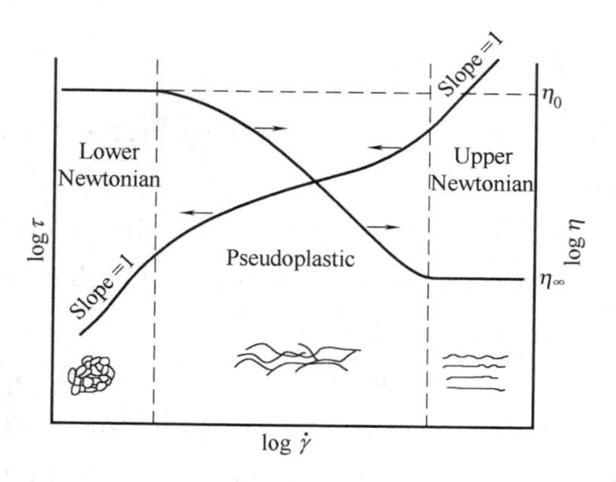

Figure 17. 1 Generalized flow properties for polymer melts and solutions

It is worthwhile to consider here what happens to the highly oriented molecules when the shear field is removed. The randomizing effect of thermal energy tends to return them to their low-shear configurations, giving rise to an elastic retraction.

The data of some actual flow curves for polymer melts cover only a portion of the general range described above, because very few instruments are capable of obtaining data over the entire range. The polyisobutylene data cover the transition from lower Newtonian to pseudoplastic regions, but the polyethylene data are confined to the pseudoplastic region. No trace of the upper Newtonian region is seen for either material at the shear rates investigated.

1. New words

logarithmic[ˌlɔgə'riθmik] adj. 对数的
rationalize[ræʃənlaiz] v. 合理地说明
random['rændəm] adj. 无规的
tendency['tendənsi] n. 趋势
untangle[ʌn'tæŋgl] v. 解缠结
orient['ɔːriənt] v. 取向
polyisobutylene[ˌpɔliaisəu'bjuːtilin] n. 聚异丁烯

2. Phrases and expressions

zero-shear viscosity 零切黏度
in terms of 用术语
thermal motion 热运动
main-chain bond 主链键
with respect to 关于

3. Notes to the text

① At low shear, the randomizing effect of the thermal motion of the chain segments overcomes any tendency toward molecular alignment in the shear field. 低剪切时,链段热运动产生的无规作用克服了剪切区域中任何使分子有序排列的趋势。

② Intense shearing eventually leads to extensive breakage of main-chain bonds, that is, mechanical degradation. 强烈的剪切作用最终导致了主链键的断链,即机械降解。

4. Exercises

(1) What differences does Figure 1 have, compared with some actual flow curves for polymer melts?

(2) What is the meaning of the word "pseudoplastic"?

(3) What is the definition of "chain segment"?

(4) Explain the two curves in Figure 1 in your own words.

(5) Translate the following sentences into Chinese:

It is worthwhile to consider here what happens to the highly oriented molecules when the shear field is removed. The randomizing effect of thermal energy tends to return them to their low-shear configurations, giving rise to an elastic retraction.

Reading Material

Creep and Stress Relaxation of Polymer

Creep Testing

Let's examine the response of the Maxwell element in two mechanical tests commonly applied to polymers. First consider a creep test, in which a constant stress is instantaneously (or at least very rapidly) applied to the material, and the resulting strain is followed as a function of time. Deformation after removal of the stress is known as *creep recovery*①.

As shown in Figure 17.2, the sudden application of stress to a Maxwell element causes an instantaneous stretching of the spring to an equilibrium value of τ_0/G, where τ_0 is the constant applied stress. The dashpot extents linearly with time with a slope of τ_0/η, and will continue to do so as long as the stress is maintained②. Thus, the *Maxwell element is a fluid*, because it will continue to deform as long as it is stressed. The creep response of a Maxwell element is therefore

$$\gamma(t)=\frac{\tau_0}{G}+\frac{\tau_0}{\eta}t \tag{17.1a}$$

or, in terms of a creep compliance, $J_c(t)\equiv\gamma(t)/\tau_0$,

$$J_c(t)\equiv\frac{\gamma(t)}{\tau_0}=\frac{1}{G}+\frac{t}{\eta} \tag{17.1b}$$

The creep compliance $J_c(t)$, being independent of the applied stress τ_0 (for a linear material), is a more general way to represent the creep response. When the stress is removed at time t_s, the spring immediately contracts by an amount equal to its original extension, a process known as *elastic recovery*. The dashpot, of course, does not recover, leaving a permanent set of $(\tau_0/\eta)t_s$, the amount the dashpot has extended during the application of stress. Although real materials never show sharp breaks in a creep test as does the Maxwell element, the Maxwell element does exhibit the phenomena of elastic strain, creep, recovery, and permanent set, which are often observed with real materials.

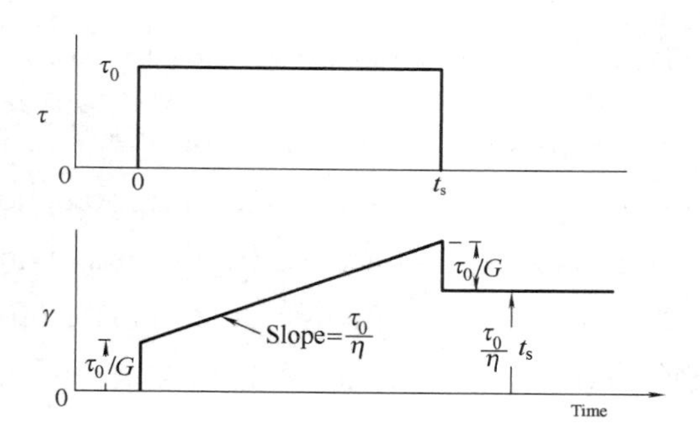

Figure 17.2 Creep response of Maxwell element

Stress Relaxation

Another important test used to study viscoelastic response is stress relaxation. A stress-relaxation test consists of suddenly applying a strain to the sample, and following the stress as a function of time as the strain is held constant. When the Maxwell element is strained instantaneously, only the spring can respond initially (for an infinite rate of strain, the resisting force in the dashpot is infinite) to a stress of $G\gamma_0$, where γ_0 is the constant applied strain. The extended spring then begins to contract, but the contraction is resisted by the dashpot. The more the spring retracts, the smaller is its restoring force, and the rate of retraction drops correspondingly③. Solution of the differential equation with $\dot{\gamma}=0$ and the initial condition $\tau=G\gamma_0$ at $t=0$ shows that the stress undergoes a first-order exponential decay:

$$\tau(t)=G\gamma_0\cdot e^{-t/\lambda} \tag{17.2a}$$

or, in terms of a relaxation modulus, $G_r(t)\equiv\tau(t)/\gamma_0$,

$$G_r(t)\equiv\frac{\tau(t)}{\gamma_0}=Ge^{-t/\lambda} \tag{17.2b}$$

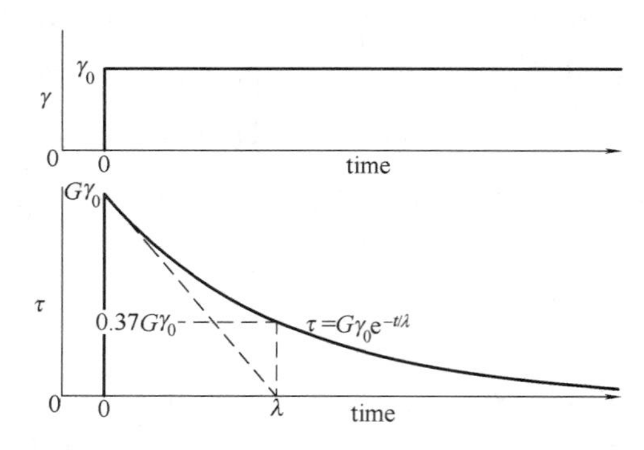

Figure 17.3 Stress relaxation of Maxwell element

Again, the relaxation modulus $G_r(t)$ is a more general means of representing stress-relaxation response because it is independent of the applied strain for linear materials④.

From (17.2), we see that the relaxation time λ is the time constant for the exponential decay, that is, the time required for the stress to decay to a factor of $1/e$ or 37% of its initial value. The stress asymptotically drops to zero as the spring approaches complete retraction (Figure 17.3).

Stress-relaxation data for linear polymers actually look like the curve for the Maxwell element. Unfortunately, they can't often be fitted quantitatively with a single value of G and a single value of λ; that is, the decay is not really first order.

1. New words

creep[kriːp] *n.* 蠕变
instantaneously[ˌinstən'teiniəsli] *adv.* 霎时,立即
strain[strein] *n.* 应变
deform[di'fɔːm] *v.* 变形
contract[kən'trækt] *v.* 收缩,缩短
dashpot['dæʃpɔt] *n.* 转折点,黏壶
exponential[ˌekspəu'nenʃəl] *adj.* 幂律的
asymptotically[əˌsaimp'tɔtikəli] *adv.* 渐近

2. Phrases and expressions

elastic recovery 弹性恢复
stress relaxation 应力松弛
creep test 蠕变测试
Maxwell element 麦克斯韦单元
restore force 回复力
relaxation modulus 松弛模量

3. Notes to the text

① ... in which a constant stress is instantaneously (or at least very rapidly) applied to the material, and the resulting strain is followed as a function of time. Deformation after removal of the stress is known as creep recovery. 在蠕变测试中,瞬时的(或者至少非常迅速地)恒定应力施加到材料上,产生的应变是时间的函数。移去应力后产生的变形就称

作蠕变回复。

② The dashpot extents linearly with time with a slope of τ_0/η, and will continue to do so as long as the stress is maintained. 转折点随时间线性发展,斜率是 τ_0/η,并且只要应力存在,这种线性发展就一直存在下去。

③ The more the spring retracts, the smaller is its restoring force, and the rate of retraction drops correspondingly. 弹簧收缩得越多,它的回复力越小,并且收缩率也相应地下降。

④ . . . the relaxation modulus $G_r(t)$ is a more general means of representing stress-relaxation response because it is independent of the applied strain for linear materials. 松弛模量 $G_r(t)$ 是一个更为普遍的表征应力松弛响应的方法,因为它是独立于施加在线性材料上的应变。

Lesson 18 Linear Viscoelasticity of Polymer

Engineers have traditionally dealt with two separate and distinct classes of materials: the viscous fluid and the elastic solid. Design procedures based on these concepts have worked pretty well because most traditional materials (water, air, steel, concrete), at least to a good approximation, fit into one of these categories. The realization has grown, however, that these categories represent only the extremes of a broad spectrum of material response①. Polymer systems fall somewhere in between, giving rise to some of the unusual properties of melts and solutions described previously. Other examples are important in the structural applications of polymers. In a common engineering stress-strain test, a sample is strained at an approximately constant rate, and the stress is measured as a function of strain. With traditional solids, the stress-strain curve is pretty much independent of the rate at which the material is strained. The stress-strain properties of many polymers are markedly *rate dependent*, however. Similarly, polymers often exhibit pronounced creep and stress relaxation②. While such behavior is exhibited by other materials (metals near their melting points, for example), at normal temperatures it is negligible, and is not usually included in design calculations. If the time-dependent behavior of polymers is ignored, the results can sometimes be disastrous.

As an aid in visualizing viscoelastic response, we introduce two linear mechanical models to represent the extremes of the mechanical response spectrum. The spring in Figure 18.1(a) represents a linear elastic or Hookean solid whose constitutive equation (relation of stress to strain and time) is simply $\tau = G\gamma$, where G is a (constant) shear modulus. Similarly, a linear viscous or Newtonian fluid is represented by a dashpot (some sort of piston moving in a cylinder of Newtonian fluid) whose constitutive equation is $\tau = \eta\dot{\gamma}$, where η is a (constant) viscosity. The strain is represented by the extension (stretching) of the model.

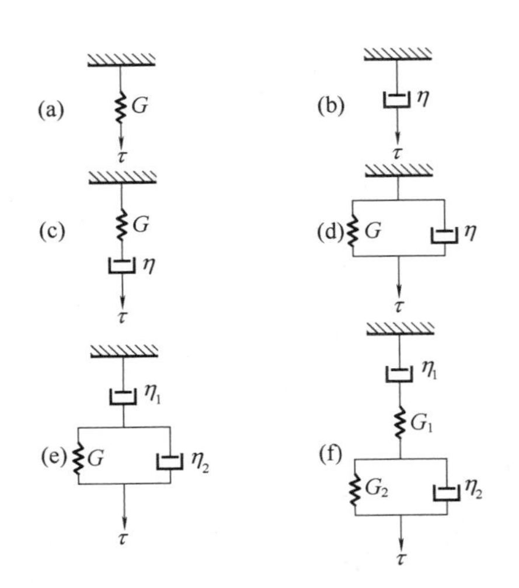

Figure 18.1 Linear viscoelastic models
(a) linear elastic (b) linear viscous
(c) Maxwell element (d) Voigt-Kelvin element
(e) three-parameter (f) four-parameter

Although the models developed here are to be visualized in tension, the notation used is for pure shear (viscometric) deformation. The equations are equally applicable to tensile deformation by replacing the shear stress τ with the tensile stress σ, the shear strain γ with the tensile strain ε, Hooke's modulus G with Young's (tensile) modulus E, and the Newtonian (shear) viscosity η with the elongational (Trouton) viscosity η_e.

Some authorities object strongly to the use of mechanical models to represent materials. They point out that real materials are not made of springs and dashpots. True, but they're not made of equations either, and it's a lot easier for most people to visualize the deformation of springs and dashpots than the solutions to equations③.

A word is needed about the meaning of the term linear. For the present, a linear response will be defined as one in which the ratio of overall stress to overall strain, the overall modulus $G(t)$, is a function of time only, not of the magnitudes of stress or strain:

$$G(t) \equiv \frac{\tau}{\gamma} = \text{function of } t \text{ only for linear response} \qquad (18.1)$$

1. New words

approximation[ə'prɔksimeiʃən] *n.* 近似值
markedly['ma:kidli] *adv.* 明显地
spring[spriŋ] *n.* 弹簧
notation[nəu'teiʃən] *n.* 记号法,表示法
object[əb'dʒekt] *v.* 反对
spectrum['spektrəm] *n.* 光,光谱 ,型谱,频谱

2. Phrases and expressions

viscous fluid　黏性流体
elastic solid　弹性固体
structural application　结构上的应用
mechanical model　力学模型
magnitude of stress or strain　应力或应变的大小

3. Notes to the text

① The realization has grown, however, that these categories represent only the extremes of a broad spectrum of material response.　然而,人们现在已经认识到,这两种类型仅仅代表各种各样的材料响应的两个极端的情况。

② The stress-strain properties of many polymers are markedly rate dependent, however. Similarly, polymers often exhibit pronounced creep and stress relaxation.　然而许多聚合物的应力—应变性质是明显依赖于比率的。同样,聚合物常常呈现出显著的蠕变和应力松弛。

③ They point out that real materials are not made of springs and dashpots. True, but they're not made of equations either, and it's a lot easier for most people to visualize the deformation of springs and dashpots than the solutions to equations.　他们指出,真实的材料并不是由弹簧和黏壶组成的。没错,但它们同样也不是由公式组成的,并且对于大多数人来说,通过弹簧和黏壶使形变形象化,较公式的解答更简单些。

4. Exercises

(1) What is the difference between polymer and traditional solid in a common engi-

neering stress-strain test?

(2) Example 1 (Figure 18.1) analyzes the response of a Maxwell element in an engineering stress-strain test. Do the same thing for a Voigt-Kelvin element. Illustrate the effect of strain rate with a sketch.

(3) Consider the three-parameter model [Figure 18.1(e)].

a. Write an expression for $J(t)$ in terms of model parameters.

b. Write an expression for the characteristic time λ_c in terms of model parameters.

c. A three-parameter model initially free of stress and strain is subjected to the following stress history:

$0 \leqslant t \leqslant \lambda$ $\tau = \tau_0$

$\lambda \leqslant t \leqslant 2\lambda$ $\lambda = 2\tau_0$

$2\lambda \leqslant t$ $\tau = 0$

Sketch $\gamma(t)$ and determine the equilibrium strain $\gamma(\infty)$.

(4) Creep data for a particular material are fit well with a standard four-parameter model [Figure 18.1(f)], with the following parameters:

$G_1 = 1250\text{dyn/cm}^3$ $\eta_1 = 1.0 \times 10^6\text{P}$

$G_2 = 500\text{dyn/cm}^3$ $\eta_2 = 2.5 \times 10^5\text{P}$

For the creep response of this material, calculate

a. λ_c

b. $\gamma(1000\text{s})$ for the applied stress history.

$t \leqslant 0\text{s}$ $\tau = 0\text{dyn/cm}^2$

$0 \leqslant t \leqslant 500\text{s}$ $\tau = -1000\text{dyn/cm}^2$

$500 \leqslant \text{ts}$ $\tau = +2000\text{dyn/cm}^2$

Reading Material

Mechanical Behavior of Glassy, Amorphous Polymers Phenomenological Models

The deformation of polymers depends very strongly upon the state of order and molecular mobility. The presentation in this section is limited to the behavior accompanying small-strain deformations, of the order of 1% or less. The strain (ε) is defined as $\varepsilon = \Delta L/L_0$, where ΔL is the stress-induced increase of specimen length and L_0 is the initial specimen length. Glassy amorphous polymers tend to deform in an approximately elastic manner and so also do crosslinked amorphous polymers above their glass transition temperature①. Uncrosslinked amorphous polymers deform viscously at temperatures above the glass transition temperature. At temperatures between the glass transition temperature and the temperature of the subglass relaxation, amorphous polymers are an elastic. An elastic polymers creep upon loading but recover completely on removal of stress provided that the maximum strain is kept at small values, typically less than 1%.

The creep compliance $J(t)=\varepsilon(t)/\sigma$, where $\varepsilon(t)$ is the time-dependent strain and σ is the constant stress is, at very small strains, less than 1%, approximately independent of stress. The material is said to be linear an elastic or linear viscoelastic. Materials of this category follow the so-called Boltzmann superposition principle which can be expressed as follows:

$$\varepsilon(t) = \sum_{i=1}^{N} j(t-t_i)\Delta\sigma_i \qquad (18.2)$$

where $\varepsilon(t)$ is the strain at time t, J is the creep compliance, and t_i is the time for the application of a new incremental load $\Delta\sigma_i$; The strain response to a certain incremental applied stress is independent of previous and future applied stresses. The strain can in this way be calculated for any stress history provided that $J(t)$ is known. In this chapter we shall consider only the mechanical behavior at such small strains that the material is approximately linear an elastic.

The lack of a satisfactory molecular theory for the deformation of polymers has led to the development of mechanical analogues and phenomenological models that represent the material②. The task is to find combinations of elastic and viscous elements that reproduce the material behavior. There is more than one combination that will reproduce the same behaviour. Some combinations are more convenient for a given kind of test and less convenient for another. However, a proper combination of elements should in principle be able to represent all the various tests. A condensed summary of the stress-strain behavior of the various models is given below.

The ultimate (fracture) properties of a wholly polymer amorphous strongly dependent are on temperature. At low temperatures, in the elastic region, the fracture is predominantly brittle and the fracture toughness is low. A considerable increase in the fracture toughness accompanies the onset of the subglass process when approaching the rubbery region. Polymers in the rubber plateau region have a low stiffness and a considerable fracture strain. Network polymers in this temperature region are elastic rubbers with high extensibility.

Maxwell model

The Maxwell model consists of an elastic spring and a viscous element (dashpot) coupled in series. The stress-strain behaviour of this model in creep (constant stress) and relaxation (constant strain) is shown in Figure 18.2 and its analytical expressions are as follows :

$$\varepsilon = \frac{\sigma_0}{E} + \frac{\sigma_0}{\eta}t \quad (\text{constant stress} = \sigma_0) \tag{18.3}$$

$$\sigma = \sigma_0 \exp\left(-\frac{E \times t}{\eta}\right) = \sigma_0 \exp\left(-\frac{t}{\tau}\right) \quad (\text{constant strain}) \tag{18.4}$$

where $\tau = n/E$ is the relaxation time. The dimension of τ is seconds, but it is only to be considered as a time constant for the model. In a stress relaxation experiment (constant E), it is equal to the real time needed to lower the stress from its original value σ_0 to σ_0/e.

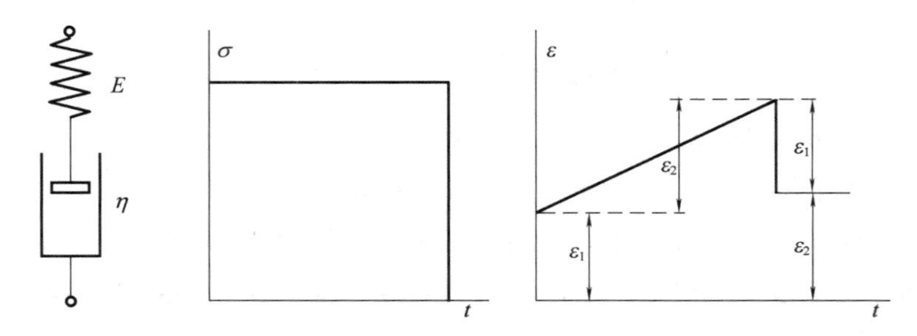

Figure 18.2 Response of a Maxwell element

Voigt-Kelvin model

The Voigt-Kelvin model consists of an elastic element and a viscous element coupled in parallel (Figure 18.3). The following constitutive equations are obtained by solving the differential equation:

$$\varepsilon = \frac{\sigma_0}{E}\left[1 - \exp\left(-\frac{E \times t}{\eta}\right)\right] = \frac{\sigma_0}{E}\left[1 - \exp\left(-\frac{t}{\tau}\right)\right] \quad (\text{constant stress} = \sigma_0) \tag{18.5}$$

where t is the retardation time. The retardation time (in seconds) is a system response time.

$$\sigma = \varepsilon_0 E \quad (\text{constant strain} = \varepsilon_0) \tag{18.6}$$

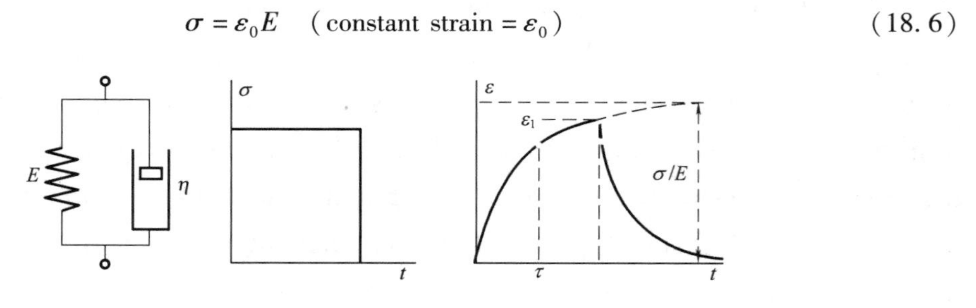

Figure 18.3 Response of a Kelvin-Voigt element

Burger's model

The Burger's model is obtained by combining the Maxwell and Voigt-Kelvin elements in series (Figure 18.4). The following equation holds under constant stress conditions:

$$\varepsilon = \frac{\sigma_0}{E_1} + \frac{\sigma_0}{E_2}t + \frac{\sigma_0}{E_3}\left[1 - \exp\left(-\frac{E_2}{\eta_2}\right)\right]$$
$$= \frac{\sigma_0}{E_1} + \frac{\sigma_0}{E_2}t + \frac{\sigma_0}{E_3}\left[1 - \exp\left(-\frac{t_2}{\tau_2}\right)\right] \quad (18.7)$$

The first term is the elastic response from spring1, the second term is the viscous flow response from dashpot I, and the final term is the response of the Voigt-Kelvin element.

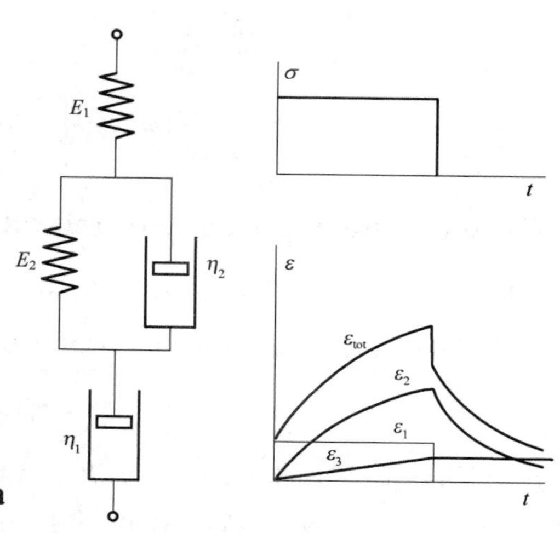

Figure 18.4 Response of a Burgers model

Dynamic mechanical behaviour of the Maxwell element and relaxation spectra

A sinusoidal strain gives a sinusoidal stress in the Maxwell element and following frequency dependence (ω is the angular velocity) is obtained for the storage modulus (E') and the loss modulus (E''):

$$E' = E\left(\frac{\omega^2\tau^2}{1 + \omega^2\tau^2}\right) \quad (18.8)$$

$$E'' = E\left(\frac{\omega\tau}{1 + \omega^2\tau^2}\right) \quad (18.9)$$

Figure 18.5 shows the storage modulus and the loss modulus as functions of the angular velocity (ω). The loss modulus passes through a maximum at $\omega = 1/\tau$. The stress-strain curves of polymers cannot be described by these simple expressions. A more complex combination of elements is required. Stress relaxation cannot be described by a single of Maxwell element alone. A number of Maxwell elements coupled in parallel are needed. The resulting relaxation modulus becomes the sum of the responses of the individual elements:

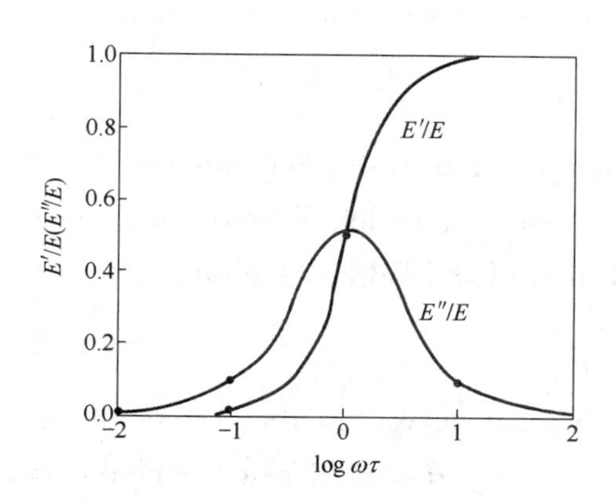

Figure 18.5 Storage and loss modulus as a function of frequency (angular velocity) for a single Maxwell element

$$E(t) = \sum_{i=1}^{n} E_i \exp\left(-\frac{t}{\tau_i}\right) \quad (18.10)$$

or, in continuous terms:

$$E(t) = E_{sp} + \int_{-\infty}^{+\infty} H(\tau)\exp\left(-\frac{t}{\tau}\right)\mathrm{d}\ln\tau \quad (18.11)$$

where E_{sp} is the relaxed modulus (time→∞) and $H(\tau)$ is the relaxation time spectrum. The components of the dynamic modulus become:

$$E'(\omega) = E_{sp} + \int_{-\infty}^{+\infty} H(\tau) \frac{\omega^2\tau^2}{1+\omega^2\tau^2} d\ln\tau \qquad (18.12)$$

$$E''(\omega) = \int_{-\infty}^{+\infty} H(\tau) \frac{\omega\tau}{1+\omega^2\tau^2} d\ln\tau \qquad (18.13)$$

Temperature dependence of relaxation time spectrum

The temperature dependence of the relaxation processes is expressed by one of two equations: the Arrhenius equation or the WLF equation. The Arrhenius equation is:

$$\tau = \tau_0 \exp\left(+\frac{\Delta E}{RT}\right) \qquad (18.14)$$

where ΔE is the activation energy, R is the gas constant and τ_0 is the pre-exponential factor (relaxation time). The Arrhenius equation can be fitted to data from isothermal plots of the loss modulus against frequency, recalling that the loss peak appears at the angular frequency:

$$\omega = \frac{1}{\tau} \Rightarrow f = \frac{1}{2\pi\tau} \qquad (18.15)$$

where f is the frequency (in hertz). A so-called Arrhenius diagram is obtained by recording the frequency (f_{max}) associated with the loss maximum at different temperature and by adapting these data to the equation:

$$\frac{1}{2\pi f} = \frac{1}{2\pi f_0}\exp\left[+\left(\frac{\Delta E}{RT}\right)\right] \Rightarrow f = f_0 \exp\left[-\left(\frac{\Delta E}{RT}\right)\right] \qquad (18.16)$$

It should be noted that the maximum of the loss peak corresponds to the central relaxation time. All real polymers exhibit a distribution in relaxation times, a relaxation time spectrum.

Taking the frequency at the loss maximum (f_{max}) for each temperature and plotting $\log f_{max}$ against 1/T, the activation energy of the loss process is obtained as follows:

$$\Delta E = -2.303R\left[\frac{d\log f_{max}}{d(1/T)}\right] \qquad (18.17)$$

The subglass relaxation processes obey Arrhenius temperature dependence.

The temperature dependence of the glass-rubber transition follows the Vogel-Fulcher equation, which is essentially a generalization of the WLF equation:

$$\tau = \tau_0 \exp\left(+\frac{c}{T-T_0}\right) \qquad (18.18)$$

Figure 18.6 illustrates the WLF behaviour appears curved in the Arrhenius plot and the curve approaches a singularity at temperature T_0. It should also be noted that the term 'activation energy' does not apply to relaxation processes showing WLF temperature dependence. In a narrow temperature region the curve may be approximated by a straight line and in that sense the activation energy may be used as a shift factor. The

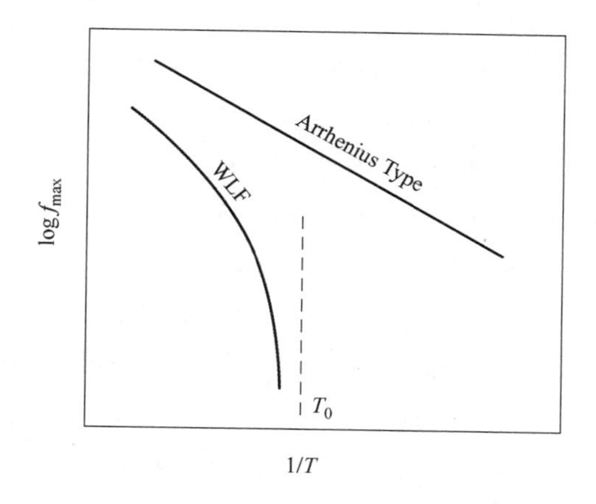

Figure 18.6 Illustration of different temperature dependences of relaxation processes

validity range of the WLF equation is from T_g to T_g + 100K.

1. New word

specimen['spesimin] *n.* 样品,样本;标本
superposition[ˌsju:pəpə'ziʃən] *n.* [数]叠加,重合
differential[difə'renʃ(ə)l] *adj.* 微分的,差别的,特异的; *n.* 微分,差别
analogues[ə'nælɔ:gəs] *n.* 类似物,类似情况;*adj.* 类似的,模拟计算机的
angular['æŋgjulə] *adj.* 有角的,角度的
generalization['dʒenrələ'zeiʃən] *n.* 概括,普遍化,一般化

2. Phrases and expressions

creep compliance 蠕变柔量
rubber plateau region 橡胶平台区
angular velocity 角速度
storage modulus 储能模量
loss modulus 损耗模量

3. Notes to the text

① Glassy amorphous polymers tend to deform in an approximately elastic manner and so also do crosslinked amorphous polymers above their glass transition temperature. 玻璃状无定形聚合物往往以近似弹性的方式发生变形,交联无定形聚合物也会在其玻璃转变温度之上发生变形。

② The lack of a satisfactory molecular theory for the deformation of polymers has led to the development of mechanical analogues and phenomenological models that represent the material. 对于聚合物变形缺乏令人满意的分子理论,这导致了可以表征聚合物变形的机械模拟和表象模型的发展。

Lesson 19 Time-Temperature Superposition of Polymer

Anyone who has ever wrestled with a cheap garden hose in cold weather appreciates the fact that polymers become stiffer and more rigid at lower temperatures, while at high temperatures they are softer and more flexible. In preceding examples, we have seen that the time scale (or frequency) of the application of stress has a similar influence on mechanical properties, short times (or high frequencies) corresponding to low temperatures and long times (low frequencies) corresponding to high temperatures. The quantitative application of this idea, time-temperature superposition, is one of the most important principles of polymer physics. It is based on the fact that the Deborah number determines quantitatively just how a viscoelastic material will behave mechanically. Changing either t_s (or ω) or λ_c can change De. The nature of the applied deformation determines t_s (or ω), while a polymer's characteristic time is a function of temperature[①]. The higher the temperature, the more thermal energy the chain segments possess, and the more rapidly they are able to respond, lowering λ_c. Thus, for example, De can be doubled by halving t_s (or doubling ω in a dynamic test) or by lowering the temperature enough to double λ_c[②]. The change in mechanical response will be the same either way, according to the time-temperature superposition principle.

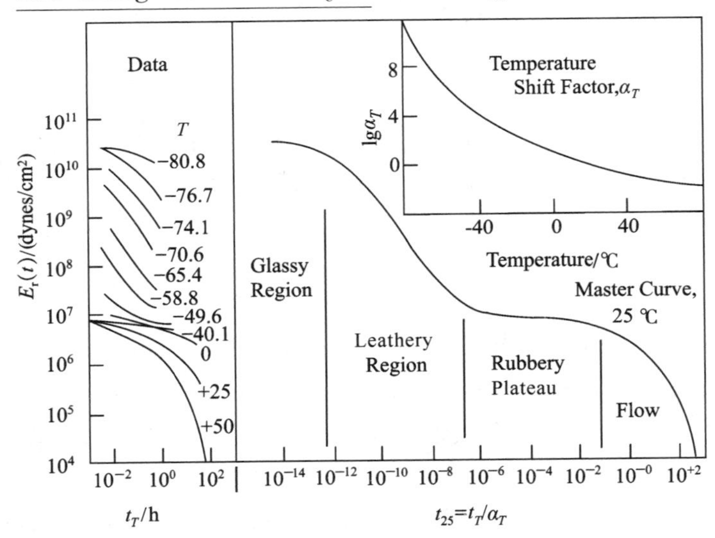

Figure 19.1 Time-temperature superposition for NBS polyisobutylene. Adapted from Tobolsky and Catsiff

Although time-temperature superposition is applicable to any viscoelastic response test (creep, dynamic, etc.), its application will be illustrated here with stress relaxation. Figure 19.1 shows tensile stress relaxation data at various temperatures for polyisobutylene, plotted in the form of a time-dependent tensile (Young's) modulus $E_r(t)$ vs. time on a log-log scale:

$$E_r(t) = \frac{\sigma(t)}{\varepsilon_0} = \frac{f(t)/A}{\Delta l/l} \tag{19.1}$$

where $f(t)$ is the measured tensile force in the sample held at a constant strain $\varepsilon_0 = \Delta l/l$ and A is its cross-sectional area. The technique is, of course, equally applicable to shear deformation. In stress relaxation, the lower measurement time limit is set by the assumption that the constant strain is applied instantaneously. In practice, inertia and other mechanical limitations make this impossible, so data are valid only at times an order of magnitude or so longer than it actually takes to apply the constant strain③. The upper limit is set by the dedication of the experimenter and the long-term stability of the sample and equipment. These data were obtained over a range from seconds to a couple of days. As might be expected, the modulus drops with time at a given temperature, and at a given time, it drops with increasing temperature.

Staring at the curves for a while indicates that they appear to be sections of one continuous curve, chopped up, with the sections displaced along the log-time axis. That this is indeed so is shown in Figure 19. 1. Here, 25℃ has arbitrarily been chosen as a reference temperature T_0 and the curves for other temperatures shifted along the log time axis to line up with it. The data below 25℃ are shifted to the left (shorter times) and those above 25℃ are shifted to the right (longer times), giving a master curve at 25℃.

1. New words

stiff[stif] *adj.* 刚性的
soft[sɔft] *adj.* 柔软的
dynamic[dai'næmik] *adj.* 动力学的
inertia[i'nə:ʃiə] *n.* 惯性,惯量
arbitrarily['a:bitrərili] *adv.* 任性地,专断地

2. Phrases and expressions

time-temperature superposition 时—温等效叠加
correspond to 相当于
polymer physics 高分子物理学
cross-sectional area 断面面积
shear deformation 剪切变形
order of magnitude 数量级

3. Notes to the text

① The nature of the applied deformation determines t_s (or ω), while a polymer's characteristic time is a function of temperature. 施加变形的本质决定着 t_s(或者 ω),而聚合物特有的时间 t_s是温度的函数。

② . . . De can be doubled by halving t_s (or doubling ω in a dynamic test) or by lowering the temperature enough to double λ_c …… De 的翻倍可以通过减半 t_s(或者在动力学测试时将 ω 翻倍),或通过将温度降低到足以使 λ_c 翻倍。

③ In practice, inertia and other mechanical limitations make this impossible, so data are valid only at times an order of magnitude or so longer than it actually takes to apply the constant strain. 实际上,惯性和其他一些机械性的限制使之成为不可能,所以数据仅在比施加恒定应变所用时间大约长一个数量级以上时才有效。

4. Exercises

Please explain time-temperature superposition.

Reading materials

Temperature Dependence of Relaxation Time Spectrum

Figures 19. 2 and 19. 3 present mechanical data showing the presence of a series of different so-called relaxation processes. It is customary to call the isochronal high-temperature process, i. e. the glass transition, α, and to give processes appearing at lower temperatures the Greek letters β, γ, δ. . . as they appear in order of descending temperature (Figure 19. 3). If we are considering a system at constant temperature, as in Figure 19. 2, the processes appear in the reverse order chronologically, i. e. in the order δ, γ, β, α.

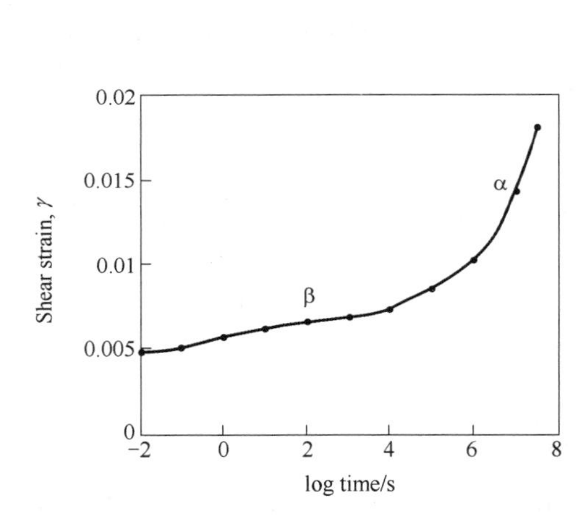

Figure 19. 2 Shear strain as a function of time (log scale) for polymethyl methacrylate (PMMA) at a constant shear stress of 7. 3MPa at 30℃. Drawn after data from Lethersich (1950)

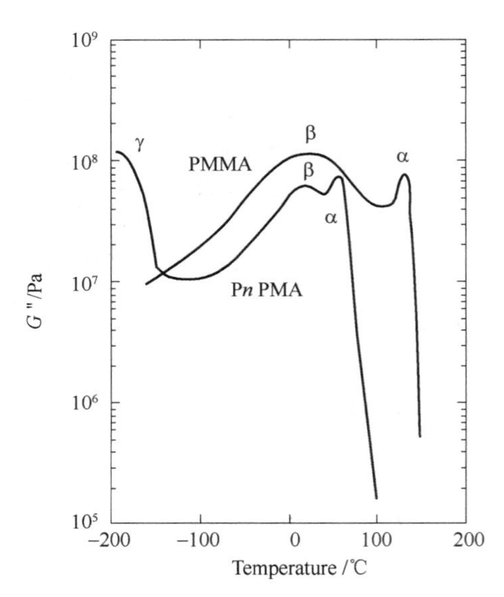

Figure 19. 3 Temperature dependence of loss modulus (G'') as a function of temperature at 1 Hz for polymethyl methacrylate (PMMA) and poly (n-propyl methacrylate) (PnPMA). Drawn after data from Heijboer (1965)

Amorphous polymers always show a glass transition process (a) and also one or more so-called *subglass* process (es), referred to as β, γ, δetc. In a constant-stress experiment, they appear as steps in the strain (creep compliance)-log time plot (Figure 19. 2). It is difficult to obtain data covering nine orders of magnitude in time as shown in Figure 19. 2. However, by taking data over a more limited time period at different temperatures, it is possible to construct a so-called master curve valid for a certain selected temperature (e. g. 30℃ as in Figure 19. 2) by predominantly horizontal shifting of the creep curves in the creep compliance-log τ diagram[①]. From data obtained at different temperatures, data for an Arrhenius diagram could be obtained by pairing log τ_{sept} and I/T. The quantity τ_{sept} refers to the time associated with the inflection point in the step in the creep compliance-log time curve.

Figure 19. 3 shows dynamic mechanical data (sinusoidally varying stress and strain) expressed in the loss modulus as a function of temperature at constant frequency. The peak maxima expressed in the pair (frequency and temperature) for each individual relaxation process can again be conveniently represented in an Arrhenius diagram.

Let us now return to the subglass processes. Their existence proves that the glassy polymer does have some limited segmental mobility which indeed is consonant with the observations of volume recovery (physical ageing) and, in general, the anelasticity If we consider the polymer at such a low temperature or for such a short time that the subglass process (es) is (are) not triggered, the material deforms elastically and shows no measurable physical ageing. Struik (1978) showed that physical ageing only appeared at temperatures between T_β (the'freezing-point' of the subglass process) and T_g. The change in modulus associated with the subglass process (es) is only small, of the order of 10% of the low-temperature, elastic value.

The molecular interpretation of the subglass relaxations has been the subject of considerable interest during the last 40 years. By varying the repeating unit structure and by studying the associated relaxation processes, it has been possible to make a group assignment of the relaxation processes. That is not to say that the actual mechanisms have been resolved. The relaxation processes can be categorized as side-chain or main-chain. Subglass processes appear both in polymers with pendant groups such as PMMA and in linear polymers such as polyethylene or poly (ethylene terephthalate). In the latter case, the subglass process must involve motions in the backbone chain.

The molecular interpretation of the relaxation processes of the methacrylates in general, and for the two shown in Figure 19. 3 in particular, is as follows.

The α process is clearly the glass transition. It is present in all polymethacrylates. It obeys WLF temperature dependence. The high-temperature subglass process β is present in all polymethacrylates. It shows both mechanical and dielectric activity and it is assigned to rotation of the side group. The β process shows Arrhenius temperature dependence.

The low-temperature subglass process (γ) is not present in PMMA and PEMA. It appears in PnPMA and longer alkyl homologues. It obeys Arrhenius temperature dependence and the activation energy is the same for all the higher polymethacrylates. It is assigned to motions in the flexible methylene sequence. It was concluded by Willbourm that a low-temperature process, essentially the same as the γ process in the higher polymethacrylates, with an activation energy of about 40kJ/mol occurred in main-chain polymers with at least four methylene groups. This low-temperature process was attributed to restricted motion of the methylene sequence, so-called crankshaft motions. The same process appears in the amorphous phase of polyethylene and is in that case also denoted γ. There are two simple conformation rearrangements that are local and that leave the surrounding stems practically unchanged. The first, suggested by Boyer(1963), involves a change from...TGT.... to...TG′ T... , i. e. a three-bond motion. The second(Schatzki crankshaft, 1963) involves the TGTGT sequence, which remains unchanged and whose surrounding bonds change conformation and cause the mid-section to rotate as crankshaft. The Schatzki crankshaft involves a considerable swept-out volume and can for that reason be excluded as a mechanism for the γ relaxation. The Boyer motion shows two energy barriers with an intermediate minimum. One of the options involves essentially an intramolecular activation energy and the swept-out volume is very small, whereas the other requires a significant swept-out volume and can for that reason be excluded. The mechanical activity i. e. the strain associated with the conformational changes, is too small for these changes to be reasonable mechanisms for the γ process. Boyd and Breitling(1974) proposed an alternative explanation closely related to the three-bond motion. He called the mechanism a left-hand-right-kink inversion. It involves the following conformational change:

$$...TTTTGTGTTTT... \rightarrow ... TTGTTTT...$$

It has a small swept-out volume and requires only a modest activation energy. The stems are slightly displaced, which leads to a change in shape (strain). Hence, the process has mechanical activity. Model calculations showed that the suggested mechanism involves activation and activation entropy(almost) similar to the experimental, but the predicted relaxation strength was significantly lower than the experimental.

Polystyrene exhibits relatively complex relaxation behavior. Apart from the glass transition(α), polystyrene exhibits four subglass relaxation processes, referred to as β, γ and δ in order of decreasing temperature. One view [McCammon, Saba and Work (1969); Sauer and Saba(1969)] is that the cryogenic process(55K in PS at 10kHz) is due to oscillatory motions of the phenyl groups, whereas others(Yano and Wada, 1971) believe that it arises from defects associated with the configuration of the polymer. The γ process appearing in PS at 180K at 10kHz has also been attributed to phenyl group oscillation or rotation. The high-temperature process denoted β occurs in PS between T_g— 100 K and T_g and is believed to be due to a rotation of the phenyl group with a main-chain cooperation.

1. New words

isochronal[ai'sɔkrənəl] *adj.* 等时线的
chronologically[ˌkrɔnə'lɔdʒikli] *adv.* 按时间的前后顺序排列地
trigger['trɪgə] *n.* 引发其他事件的一件事；*vt.* 引发，触发；
oscillation[ˌɑːsi'leiʃn] *n.* 振动；波动；动摇；<物>振荡

2. Phrases and expressions

the isochronal high-temperature process 等时的高温过程
the reverse order chronologically 倒序顺序
crankshaft motions 曲轴运动
swept-out 被卷走
activation entropy 活化熵

3. Notes to the text

① However, by taking data over a more limited time period at different temperatures, it is possible to construct a so-called master curve valid for a certain selected temperature (e. g. 30℃) by predominantly horizontal shifting of the creep curves in the creep compliance-log τ diagram. 然而，在不同的温度下有限的时间获取数据，可以构造在选定有效温度下的一个所谓的主曲线（例如30℃），主要由在蠕变柔量与松弛时间的对数图中水平移动的蠕变曲线得到。

PART 5 THERMAL ANALYSIS OF POLYMERS

Lesson 20 Thermo-analytical Methods

According to the definition originally proposed in 1969 by the Nomenclature Committee of the International Confederation for Thermal Analysis (ICTA) and later reaffirmed in 1978, thermal analysis includes a group of analytical methods by which a physical property of a substance is measured as a function of temperature while the substance is subjected to a controlled temperature regime. Thus, thermal analysis involves a physical measurement. Figure 20. 1 presents a summary of the different thermal analytical methods available. In this chapter, calorimetry, i. e. differential scanning calorimetry (DSC), and differential thermal analysis (DTA), thermogravimetry (TG), thermal mechanical analysis (TMA, DMTA), thermal optical analysis (TOA) and dielectric thermal analysis (DETA) are discussed. In the first part, the methods are briefly described. The use of thermal analysis on polymers requires special attention, as is discussed in the final section of the chapter. Examples from the melting and crystallization of flexible-chain polymers, the glass transition of amorphous polymers, phase transitions in liquid-crystalline polymers and chemical reactions including the degradation of polymers are presented to illustrate the nonequilibrium effects which are typical of polymers①.

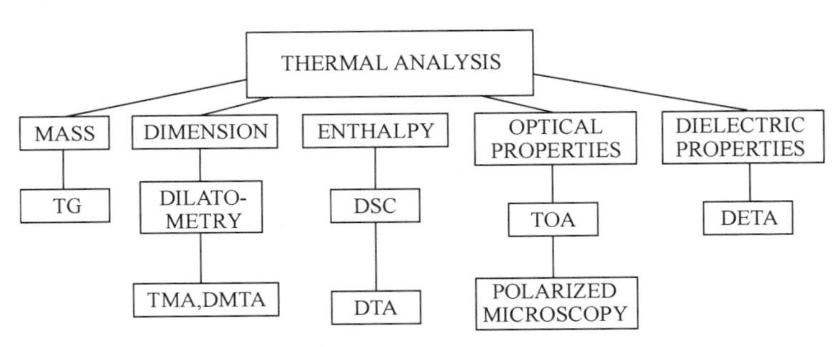

Figure 20. 1 Thermo-analytical methods

Thermo-analytical methods are powerful tools in the hands of the polymer scientist. Thermometry is the simplest and oldest method in thermal analysis. A sample is heated by a constant heat flow rate. Any phase transition is recorded as an invariance in temperature. The number of phenomena which can be directly studied by thermal analysis [DSC (DTA), TG, TMA, DMTA, TOA and DETA] is impressive. Typical of these methods is that only small amounts of sample(a few milligrams)are required for the analysis. Calorimetric methods record exo-and endothermic processes, e. g. melting, crystallization, liquid-crystalline phase transitions, and chemical reactions, e. g. polymerization, curing, depolymerization and degradation. Second-order transitions, e. g. glass transitions, are readily revealed by the calorimetric methods. Thermodynamic quantities, e. g. specific heat, are sensitively determined. TG is a valuable tool for the determination of the content of volatile species and fillers in polymeric materials and also for studies of polymer degradation②. The majority of the aforementioned physical transitions can also be monitored by TMA(dilatometry). DMTA and DETA provide information about relaxation processes, both the glass transition and the secondary transitions(subglass processes). Additional information about the nature of the phase transitions in both crystalline and liquid-crystalline polymers may be obtained by TOA.

1. New words

calorimetry[ˌkælə'rimitri] *n.* [热]量热学,热量测定

thermogravimetry[ˌθɜːməugrə'vi1mi1tri1] *n.* [分化]热重量分析法

dielectric[ˌdaii'lektrik] *adj.* 非传导性的,诱电性的;*n.* 介电,绝缘体

thermometry[θə'mɔmitri] *n.* 温度测量,温度测定法,[物]计温学

2. Phrases and expressions

differential scanning calorimetry 差示扫描量热法

differential thermal analysis 示差热分析

thermal mechanical analysis 热机械分析

thermal optical analysis 热光学分析

dielectric thermal analysis 热介电分析

3. Notes to the text

① Examples from the melting and crystallization of flexible-chain polymers, the glass transition of amorphous polymers, phase transitions in liquid-crystalline polymers and chemical reactions including the degradation of polymers are presented to illustrate the nonequilibrium effects which are typical of polymers. 以柔性链聚合物的熔融结晶、非晶聚合物的玻璃化转变、液晶聚合物的相变和包括聚合物降解在内的化学反应为例,说明了聚合物的非平衡效应。

② TG is a valuable tool for the determination of the content of volatile species and fillers in polymeric materials and also for studies of polymer degradation. TG是测定高分子材料中挥发性物质和填料含量的重要工具,也是高分子降解研究的重要工具。

4. Exercises

Demonstrate the different thermal analytical methods.

Reading Material

Thermal Optical Analysis (TOA)

TOA is normally carried out using a polarized light microscope(crossed polarizers) equipped with a hot-stage by which the temperature of the sample can be controlled. The microscopic image of the typically 10μm thick sample can be viewed directly in binoculars or the transmitted light intensity can be recorded with a photodiode①. Melting(crystallization) and mesomorphic phase transitions in liquid crystalline polymers can be directly studied by TOA(Figure 20. 2). Apparatus for TOA of sheared polymer melts was developed during the 1980s.

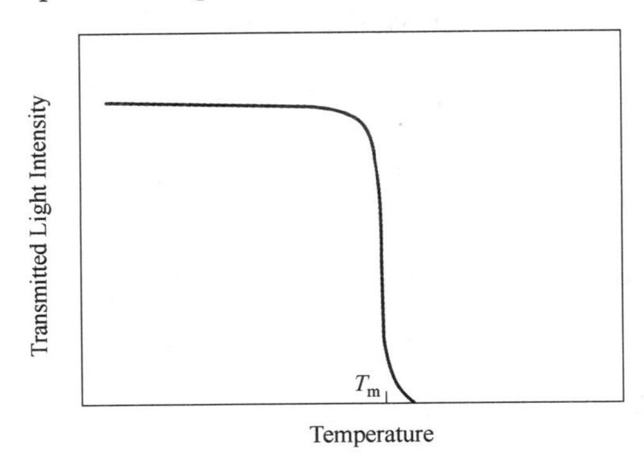

Figure 20. 2 Melting of a crystalline polymer as recorded by TOA

1. New words

binocular[bi'nɔkjulə] *adj.* [生物]双眼的,双目并用的;*n.* 双筒望远镜

photodiode[fəutəu'daiəud] *n.* [电子]光电二极管

mesomorphic[ˌmesəu'mɔːfik] *adj.* 具有中间相的

apparatus[ˌæpə'reitəs] *n.* 装置,设备,仪器,器官

2. Phrases and expressions

crossed polarizer　正交偏光镜

hot-stage　(显微镜)热台

3. Notes to the text

① The microscopic image of the typically 10μm thick sample can be viewed directly in binoculars or the transmitted light intensity can be recorded with a photodiode. 通常10μm厚样品的显微图像可以用双筒望远镜直接观察,也可以用光电二极管记录透射光强。

Lesson 21 Differential Thermal Analysis and Calorimetry

It is not the purpose of this chapter to give full details of the thermoanalytical methods. A very brief survey is, however, necessary in order to provide a basis for the understanding of the later discussion of the phenomena in relation to polymers.

Accurate temperature-measuring devices-thermocouple, resistance thermometer and optical pyrometer-were developed in the late 1800s. These instruments were used by Le Chatelier in the late nineteenth century on chemical systems to study curves of change in the heating rate of clay (see Le Chatelier 1887). The first DTA method was conceived by the English metallurgist Roberts-Austen (1889). The modem DTA instrumentation was introduced by Stone (1951). This apparatus permitted the flow of gas or vapour through the sample during the temperature scans①. The undesirable feature of the classical DTA [Figure 21. 1 (b)] in sensor-sample interaction was overcome by Boersma (1955). This type of technique has since been referred to as 'Boersma DTA' [Figure 21. 1 (c)]. The temperature sensors are placed outside the sample and the reference.

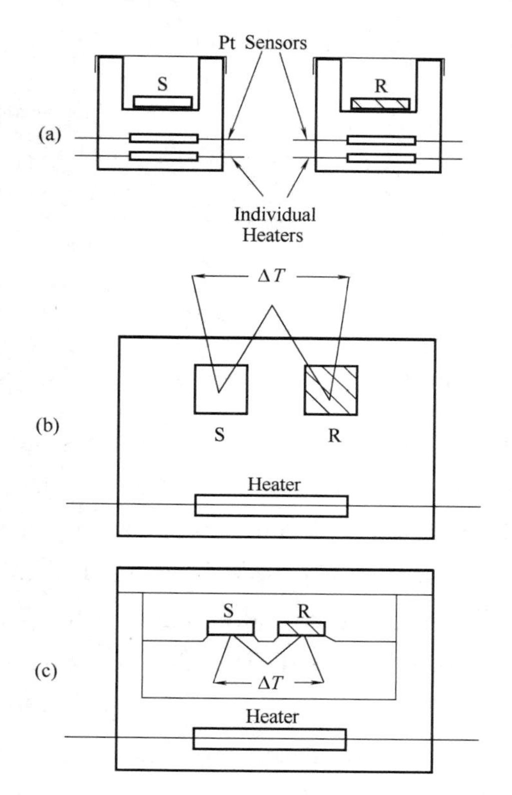

Figure 21. 1 Schematic representation of (a) DSC; (b) 'classical' DTA and (c) 'Boersma' DTA. S and R denote sample and reference, respectively

The first differential scanning calorimeter was introduced by Watson et al. (1964) [Figure 21. 1 (a)]. A number of new developments in the instrumentation have since been made. The temperature scan is controlled, and data are collected and analysed by computers in today's instruments. Simultaneous measurements of differential temperature (ΔT) and sample weight, i. e. combined DTA and TG as well as combined DTA and TOA are now commercially available. High-pressure DTA instruments have been in use since the early 1970s. In 1966, Cohen and co-workers constructed a DTA cell which could be used at pressures up to 5 MPa. DSC instruments equipped with a UV cell were developed during the 1980s.

A new class of highly sensitive (100 nW) isothermal or slowly scanning calorimeters to be used at temperatures lower than 120℃ were developed by Suurkuusk and Wadso starting in the late 1960s (see Suurkuusk and Wadso 1982). The first instrument in the series was a batch reaction calorimeter. Similar systems had been built a few years earlier by Calvet and Prat (1963) and by Benzinger and Kitzinger (1963). These thermopile heat conduction calorimeters consist of a calorimetric vessel surrounded by thermopiles, often Peltier elements through which the heat is conducted to or from the surrounding heat sink②.

The calorimetric methods, DSC and DTA are schematically presented in Figure 21. 1. DSC relies on the so-called 'null-balance' principle (Figure 21. 2). The temperature of the sample holder is kept the same as that of the reference holder by continuous and automatic adjustment of the heater power. The sample and reference holders are individually heated. A signal proportional to the difference between the heat power input to the sample and to the reference, dH/dt, is recorded. In the classical and Boersma DTA systems, the sample and reference are heated by a single heat source. Temperatures are measured by sensors embedded in the sample and reference material (classical) or attached to the pans which contain the material (Boersma). DTA measures the difference in temperature between sample and reference, but it is possible to convert dH/dt into absorbed or evolved heat via a mathematical procedure. The conversion factor is temperature-dependent. However, a DTA which accurately measures calorimetric properties is referred to as a differential scanning calorimeter. A DSC is thus a DTA that provides calorimetric glass transition information. The DSC instrument made by PerkinElmer is more a 'true' calorimeter since it directly measures differences in heat between an inert reference and the sample.

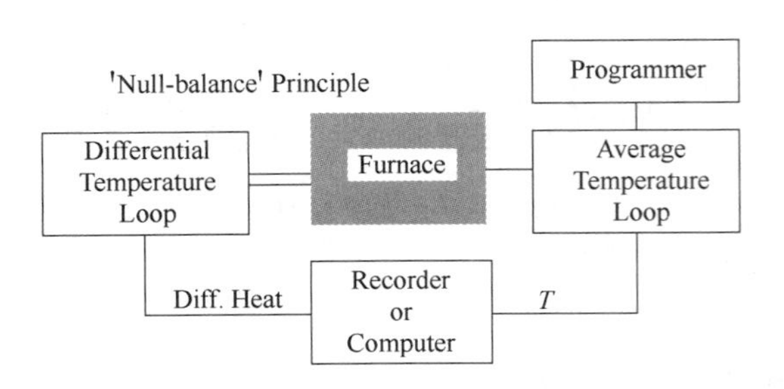

Figure 21. 2 Principle of perkin-elmer DSC apparatus

Figure 21. 3, showing the DSC thermogram of an undercooled, potentially semicrystalline polymer, illustrates the measurement principle. At low temperatures, the sample and the reference are at the same temperature (balance). When the glass transition is reached, an increase in the (endothermal) heat flow to the sample is required in order to main-

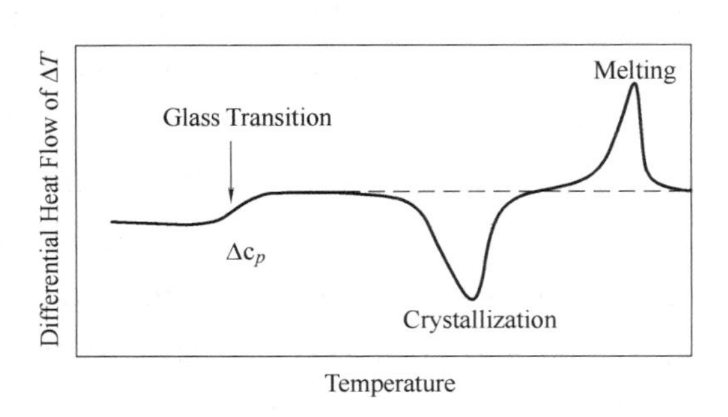

Figure 21. 3 Schematic DSC traces showing three transition types

tain the two at the same temperature. The change in level of the scanning curve is thus proportional to Δc_p. The polymer crystallizes at a higher temperature and exothermal energy is evolved. The heat flow to the sample should in this temperature region be less than the heat flow to the reference. The integrated difference between the two, i. e. the area under the exothermal peak is thus equal to the crystallization enthalpy. At further higher temperatures, melting, which is an endothermal process, occurs. The heat flow to the sample is higher than that to the reference, and the peak points upwards. The area under the endothermal peak is thus proportional to the melting enthalpy.

There are a few significant differences between a DSC (Perkin-Elmer type) and a DTA. The mass of the sample and reference holders in the DSC apparatus is very low and the maximum cooling rate is greater in the typical (Perkin-Elmer) DSC apparatus than in DTA. The maximum measurement temperature for a DSC is only about 725℃. The upper temperature limit for a number of DTA instruments is significantly greater. This is not, however, a crucial question for organic polymers. One obstacle to calorimetric operation remains unsolved for the DTA method: the factor converting the observed peak area to energy is temperature-dependent. This is particularly relevant for polymers which typically melt over a wide temperature range. The melting curve is asymmetric and requires the use of a complex conversion factor③.

The calibration constant in DSC is independent of temperature and quantitative operation is inherently simpler than with DTA.

The accuracy in the determination of transition temperatures by DSC/DTA is dependent on several factors:

· Standardized sample geometry and mass. The sample should be flat and have good thermal contact with the sample pan. Heat-conductive, 'thermally inert' liquid media may be used to improve the thermal contact.

· The purge gas and the sample pan material should be 'inert'.

· Thermal lag (difference) between sample and thermometer may be corrected for by using the slope of the leading edge of the melting of highly pure indium (or similar metal).

· Parallel processes should be inhibited. Melting of polymer crystals is accompanied by crystal thickening (parallel process).

The accuracy in the determination of transition enthalpies depends on the level of the enthalpy change, on the temperature region of the transition and on the linearity (or control) of the base line.

1. New words

pyrometer[pai'rɔmitə] n. 高温计

thermopile['θɜːmə(u)pail] n. [热]热电

电偶,[电]温差电偶

endothermal[ɛndo'θərməl] *adj.* 吸热的

exothermal[ˌeksəu'θə：məl] *adj.* 放热的,放能的

enthalpy[en'θælpi] *n.* [热]焓,[热]热函,热含量

2. Phrases and expressions

null-balance 零位平衡

enthalpy change 热含量变化

3. Notes to the text

① This apparatus permitted the flow of gas or vapour through the sample during the temperature scans. 在温度扫描过程中,这种装置允许气体或蒸气流过样品。

② These thermopile heat conduction calorimeters consist of a calorimetric vessel surrounded by thermopiles, often Peltier elements through which the heat is conducted to or from the surrounding heat sink. 这些热电堆热传导量热计由一个由热电堆包围的量热容器组成,这些热电堆通常是 Peltier 元件,热量通过这些热电堆或从周围的散热器传递。

③ One obstacle to calorimetric operation remains unsolved for the DTA method: the factor converting the observed peak area to energy is temperature-dependent. This is particularly relevant for polymers which typically melt over a wide temperature range. The melting curve is asymmetric and requires the use of a complex conversion factor. 对于 DTA 方法来说,热量计算的一个障碍仍然没有得到解决:将观测到的峰值区域转换成能量的因素是依赖温度的。这对于聚合物尤其适用,因为聚合物通常在较宽的温度范围内熔化。熔融曲线是不对称的,需要使用复杂的转换系数。

4. Exercises

(1) What kind of transition types of PE in DSC traces ?

(2) Demonstrate the effect factors on the determination of transition temperatures by DSC/DTA.

Reading Material

Thermogravimetry

The fundamental components of TG have existed for thousands of years. Mastabas or tombs in ancient Egypt (2500 BC) have wall carvings and paintings displaying both the balance and the fire. The two components were, however, first coupled in the fourteenth century AD for studies of gold refining. Honda pioneered the modem TG analytical technique (thermobalance) in 1915. The automatic thermobalance was introduced by Cahn and Schultz (1963).

TG is carried out in a so-called thermobalance which is an instrument permitting the continuous measurement of sample weight as a function of temperature/time. The following components are included in a typical TG instrument: recording balance; furnace; furnace temperature controller; and computer. Figure 21.4 shows a schematic TG trace for a filled polymer. The furnaces can be run at temperatures up to 2400℃ or more in a great variety of atmospheres, including corrosive gases. The sensitivity of the best modem recording balances is extremely high, in the microgram range.

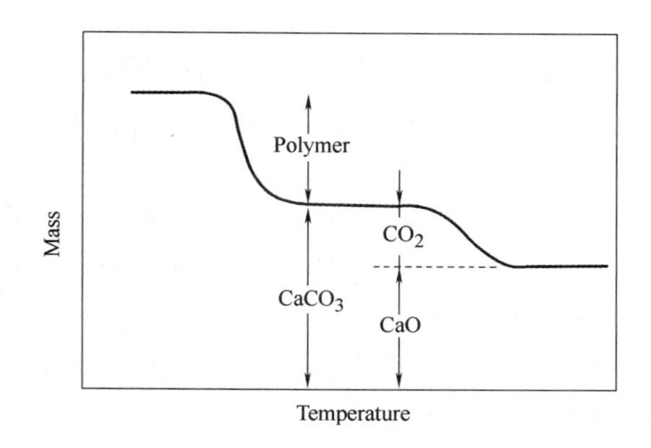

Figure 21.4 Sample mass as a function of temperature. Schematic curve of a polymer filled with $CaCO_3$

It should be noted that the sample size and form affect the shape of the TG curve. A large sample may develop thermal gradients within the sample, a temperature deviation from the set temperature due to endo-or exothermal reactions and a delay in mass loss due to diffusion obstacles①. Finely ground samples are preferred in quantitative analysis for the aforementioned reasons.

TG is often combined with various techniques to analyse the evolved gas. Infrared (IR) spectroscopy and gas chromatography (GC), the latter often combined with mass spectrometry (MS) or IR, are used for identification of the volatile products.

1. New words

thermobalance [ˌθəːməu'bæləns] *n.* [分化]热天平

corrosive [kə'rəusiv] *adj.* 腐蚀的，侵蚀性的

volatile ['vɔlətail] *adj.* [化学]挥发性的，不稳定的；*n.* 挥发物

chromatography [ˌkrəumə'tɔgrəfi] *n.* 色层分析，色谱分析法

2. Phrases and expressions

Infrared spectroscopy 红外光谱分析

gas chromatography 气相色谱分析

mass spectrometry 质谱分析法

3. Notes to the text

① A large sample may develop thermal gradients within the sample, a temperature deviation from the set temperature due to endo-or exothermal reactions and a delay in mass loss due to diffusion obstacles. 大试样内部可能出现热梯度，由于内、放热反应导致的温度偏离设定温度，以及由于扩散障碍导致的质量损失延迟。

Lesson 22 Dilatometry/Thermal Mechanical Analysis (TMA) and Dynamic Mechanical Thermal Analysis (DMTA)

The classical way of measuring sample volume as a function of temperature is by a Bekkedahl dilatometer. More recent types of instrumentation, e. g. the thermal mechanical analysers, have been developed by several companies. These instruments not only measure volume and linear thermal expansion coefficients but also modulus as a function of temperature.

When thermal expansion or penetration (modulus) is being measured, the sample is placed on a platform of a quartz sample tube. The thermal expansion coefficient of quartz is small (about $0.6 \times 10^{-6}\ K^{-1}$) compared to the polymer materials. The quartz tube is connected to the armature of a linear variable differential transformer (LVDT) and any change in the position of the core of the LVDT, which floats frictionless within the transformer coil, results in a linear change in the output voltage. The upper temperature limit for the currently available commercial instruments is about 725℃. Gillen (1978) showed that the tensile compliance as obtained by TMA using the penetration probe and suitable compressive loads is comparable with the data measured by conventional techniques on considerably larger samples.

DMTA measures stress and strain in a periodically deformed sample at different loading frequencies and temperatures. It provides information about relaxation processes in polymers, specifically the glass transition and subglass processes. It should be noted that the strains involved in the measurements should be small (less than 0.5%) to avoid a nonlinear response, i. e. nonlinear viscoelasticity. There are essentially three main types of instrument: the torsion pendulum. An apparatus based on the resonance method, and forced oscillation instruments (Figure 22.1).

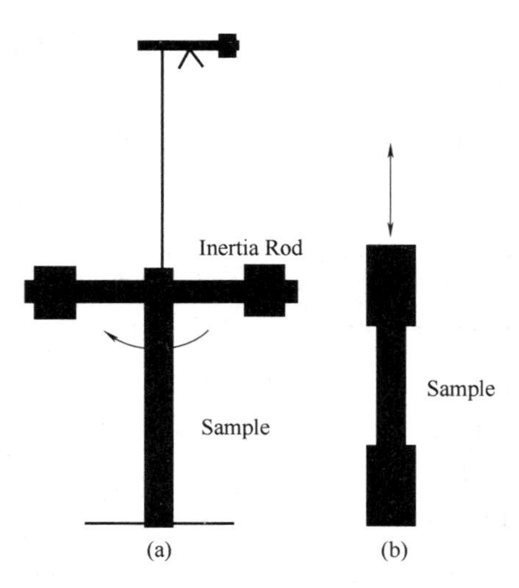

Figure 22.1 Schematic representation of (a) torsion pendulum and (b) reversed uniaxial tension (forced oscillation)

A highly schematic representation of the torsion pendulum is given in Figure 22.1 (a). The sample, which may be a cylindrical rod, is rigidly held at one end, and at the other end supports an inertia rod. The inertia rod is set into oscillation and the polymer sample is subjected to a sinusoidal torsion which, depending on the relative size of the viscous component, gradually dampens①. The frequency range of operation is 0.01 ~ 50Hz.

The forced oscillation technique can be used over a wider frequency range than the torsion pendulum ranging from 10^{-4} to 10^4 Hz. Uniaxial extension[Figure 22.1(b)], bending, torsion and shear are used in different commercial instruments. A sinusoidal strain is applied to the specimen and the stress(force) is accurately measured using a strain gauge transducer as a function of time②. Knowing both the strain and stress as functions of time enables the complex modulus to be determined.

1. New words

dilatometer[ˌdiləˈtɔmitə] *n.* [仪]膨胀计

quartz[kwɔːts] *n.* 石英

armature? [ˈɑːmətʃə] *n.* 电枢(电机的部件),盔甲

pendulum[ˌpendjuləm] *n.* 钟摆,摇锤

cylindrical[siˈlindrikəl] *adj.* 圆柱形的,圆柱体的

oscillation[ˌɔsiˈleiʃn] *n.* 振荡,振动,摆动

gauge[geidʒ] *n.* 计量器,标准尺寸; *vt.* 测量

2. Phrases and expressions

thermal expansion coefficient 热膨胀系数

compressive load 压缩负荷

torsion pendulum 扭转摆

relaxation process 弛豫过程

3. Notes to the text

① The inertia rod is set into oscillation and the polymer sample is subjected to a sinusoidal torsion which, depending on the relative size of the viscous component, gradually dampens. 将惯性杆设为振荡,聚合物试样受正弦扭转,根据黏性组分的相对大小逐渐衰减。

② A sinusoidal strain is applied to the specimen and the stress(force) is accurately measured using a strain gauge transducer as a function of time. 在试件上施加正弦应变,用应变传感器作为时间函数精确测量应力(力)。

4. Exercises

(1) Demonstrate the applications of TMA and DMTA respectively.

(2) How to measure stress and strain in a periodically deformed sample at different loading frequencies and temperatures.

Reading Material

Dielectric Thermal Analysis (DETA)

DETA, which is also referred to as dielectric spectroscopy, provides information about the segmental mobility of a polymer. Chemical bonds between unlike atoms possess permanent electrical dipole moments. Many polymers with significant bond dipole moments show no molecular dipole moments due to symmetry, i. e. the bond moments of a given central atom (or group of atoms) counteract each other and the vector sum of the dipole moments is zero①. Other polymers, however, have repeating unit structures such that the dipole moments can *vectorially* accumulate into a repeating unit moment in the possible conformational states. This group of polymers can be studied by DETA. Polarizability (α) is the sum of dipole moments per unit volume. Figure 22.2 shows the variation of α as a function of the frequency of the alternating electric field. The electric field induces a distortion of the electronic clouds at a frequency of about 1015Hz, which corresponds to the optical ultraviolet range②. Electronic polarizability (α) is closely related to the refractive index (n) of visible light. So-called atomic polarization arises from small displacements of atoms under the influence of the electric field at a frequency of approximately 10^{13} Hz (optical infrared range).

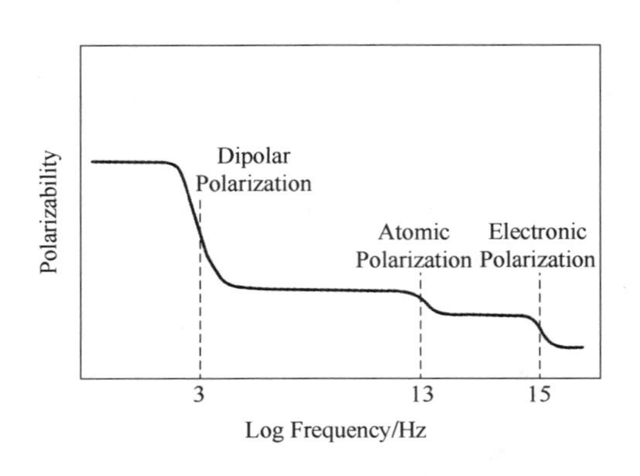

Figure 22.2 Schematic representation of polarizability as a function of frequency

The atomic polarizability cannot be determined directly but is normally small compared to the electronic polarizability. Electronic and atomic polarization occur in all types of polymer, even polymers with no permanent dipole moments. Polymers with permanent dipoles show no macroscopic polarization in the absence of an external electric field. If an alternative electric field is applied and if the electric field frequency is sufficiently low with reference to the jump frequency of segments of the polymer, the dipoles orient in the field and the sample shows not only electronic and atomic polarization but also a dipolar polarization. DETA, which operates in a frequency range from 10 to 10^8 Hz, is used to monitor dipole reorientation induced by conformational changes. These are referred to as dielectric relaxation processes.

1. New words

vector['vektə] *n.* 矢量,带菌者,航线

polarizability[ˌpəuləˌraizə'biləti] *n.* [电子]极化性,[电磁]极化度

polarization[ˌpəulərai'zeiʃən] *n.* 极化,偏振,两极分化

2. Phrases and expressions

dielectric spectroscopy 介电谱

dipole moment 偶极矩

3. Notes to the text

① Many polymers with significant bond dipole moments show no molecular dipole moments due to symmetry, i. e. the bond moments of a given central atom (or group of atoms) counteract each other and the vector sum of the dipole moments is zero. 许多具有重要键偶极矩的聚合物由于对称而没有分子偶极矩,即给定中心原子(或原子群)的键矩相互抵消,偶极矩的矢量和为零。

② The electric field induces a distortion of the electronic clouds at a frequency of about 10^{15} Hz, which corresponds to the optical ultraviolet range. 电场引起电子云的畸变频率约为 10^{15}Hz,对应于光紫外范围。

Lesson 23 Thermal Behavior of Semicrystalline Polymers

The melting of polymer crystals exhibits many instructive features of non-equilibrium behaviour. It has been known since the 1950s that the crystals of flexible-chain polymers, e. g. polyethylene(PE), are lamella-shaped with the chain axis almost parallel to the normal of the lamella. The lamellar thickness (L_c) is of the order of 10nm, corresponding to approximately 100main chain atoms, which is considerably less than the total length of the typical polymer chain. This fact led to the postulate that the macroconformation of the chains must be folded.

For linear polyethylene(LPE), the surface free energies are equal respectively to σ = 93mJ/m^2 and σ_L = 14mJ/m^2, resulting in an equilibrium value(L_c/B) of 6.6 which is three orders of magnitude greater than the experimental value. The crystal thickening of polymer crystals expected to occur on the basis of these thermodynamic arguments has been verified by X-ray diffraction experiments①. The recorded dependence of melting point on heating rate of single crystals of linear polyethylene presented in Figure 23.1 is consonant with this view. At low heating rates, the crystal thickening occurs to a much greater extent than at high heating rates, which in turn leads to a greater 'final' crystal thickness and a higher melting point after slow heating. The melting point value obtained at the higher heating rates is thus more in agreement with that of the original crystals.

Extended-chain crystals of polyethylene are produced by high-pressure crystallization at elevated temperatures, typically at 0.5MPa and 245℃. These micrometre-thick crystals display a distinctly different melting behaviour from that of the thin folded-chain single crystals grown from solution②. The recorded increase in melting point with increasing heating rate shown in Figure 23.2 is due to superheating.

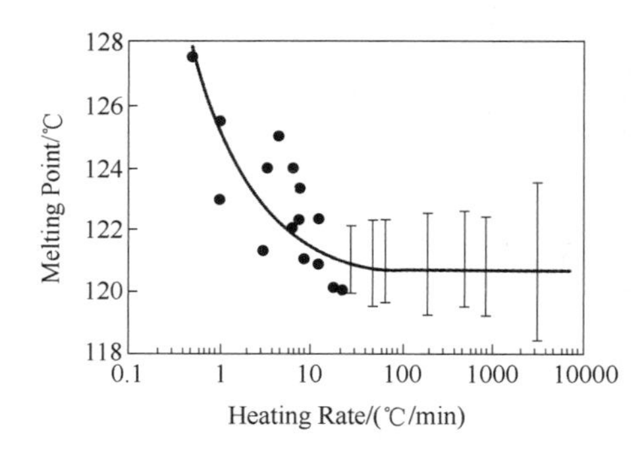

Figure 23.1 Melting point of solution crystals of LPE [0.05% (w/w) in toluene at 81℃] as a function of heating rate. Drawn after data from Hellmuth and Wunderlich(1965)

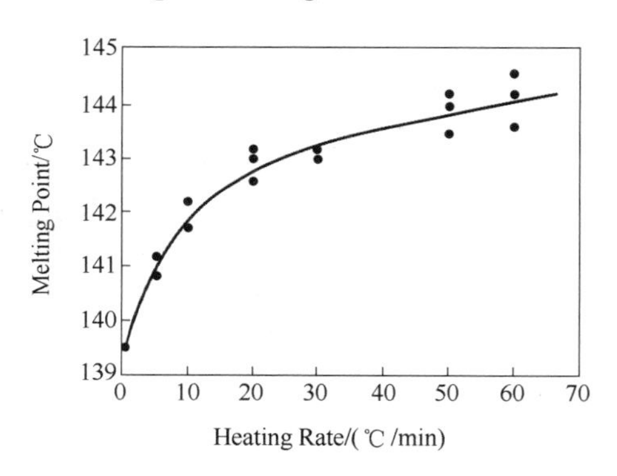

Figure 23.2 Melting point of extended-chain crystals of linear polyethylene(0.48MPa; 227℃). Drawn after data from Hellmuth and Wunderlich(1965)

Crystals of intermediate thickness display approximately zero-entropy-production melting, i. e. the melting point is almost independent of heating rate. Interestingly, as is shown in Figure 23. 3, linear polyethylene samples with a broad molar mass distribution contain crystals with a great variety of thicknesses, displaying in a single sample reorganization, zero-entropy-production melting and superheating in order of increasing crystal thickness, i. e. of increasing melting point. Polymers in general melt over a wide temperature range, typically covering more than 30℃. This is due, first, to their multicomponent nature. Polymers always exhibit a distribution in molar mass and occasionally also in monomer sequence (copolymers). The different molecular species crystallize at different temperatures and this leads to a significant variation in crystal thickness and melting point. Polymer samples which have first crystallized under isothermal conditions and have later been rapidly cooled to lower temperatures frequently exhibit bimodal melting for these reasons. The high-temperature peak is associated with the high molar mass species which have crystallized under the isothermal conditions and the low-temperature peak is due to the low molar mass component able to crystallize only during the subsequent cooling phase.

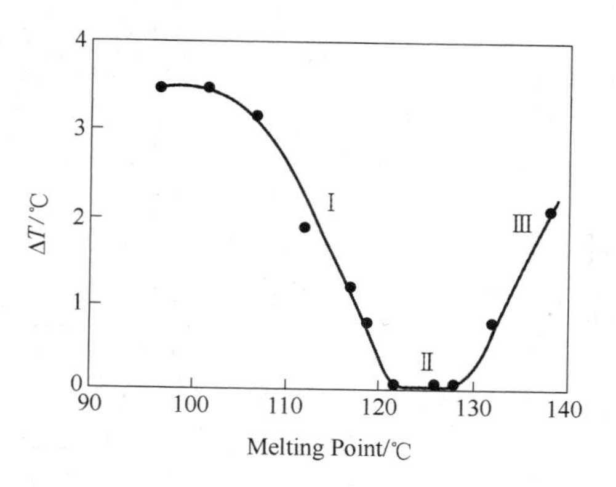

Figure 23. 3 Difference (AT) in melting point as recorded at 10℃/min from that recorded at high heating rates (region Ⅰ: crystal thickening) or zero heating rate (region Ⅲ: superheating) plotted against melting point as recorded at 10℃/min. Region Ⅱ refers to zero-entropy-production melting. Drawn after data from Gedde and Jansson (1983) on linear polyethylene

Second, reorganization of crystals may occur on heating, involving either partial melting followed by crystallization forming thicker and more perfect crystals of essentially the same unit cell of the original crystals or the transformation of one crystal structure to another. Samples which undergo these premelting transitional phenomena display bimodal or multimodal melting provided that the original crystals are of about the same lamella thickness③.

Figure 23. 4 presents four cases of bimodal melting with different reasons for the bimodality. The melting of the once-folded orthorhombic crystals of n-$C_{294}H_{590}$ leads to recrystallization into extended-chain crystals which melt at a higher temperature in the high-temperature melting peak④. Isothermal crystallization (at T_c' Figure 23. 4) almost always leads to two melting peaks with a minimum in between at a temperature in the vicinity of T_c. High-temperature annealing leads to similar effects. Immiscible, not co-

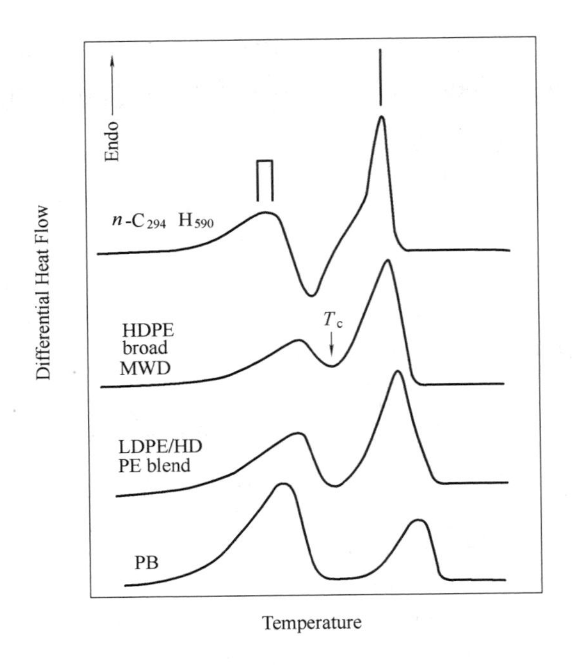

Figure 23.4 Heating thermograms showing different bimodal melting traces. Schematic curves

crystallizing binary blends of polymers with different melting points show two melting peaks. Numerous polymers have different possible crystalline structures, i. e. they have different polymorphs, each with a certain melting point. The content of the crystalline component, the crystallinity, in semicrystalline polymers is a major factor affecting their material properties, e. g. modulus, permeability and density. The crystallinity of a sample can be determined by several techniques, e. g. X-ray diffraction, density measurements and DSC/DTA.

The mass crystallinity(ω_c) obtained from the heat of fusion is based on the measurement of the area under the DSC melting peak. The choice of base line is crucial particularly for polymers of low crystallinity, e. g. poly(ethylene terephthalate)(PET). Another problem arises from the fact that the heat of fusion is temperature-dependent. What temperature should be selected? Two rigorous methods have been proposed by Gray(1970) and Richardson(1976): the total enthalpy method and the peak area method.

Excellent agreement is obtained for crystallinity data obtained for samples of linear, branched and chlorinated polyethylene and PETP by the DSC and total enthalpy methods with those obtained by X-ray diffraction method. Numerous specific heat data on different polymers were collected in Wunderlich and Baur(1970).

The greatest problem is to determine the crystallinity of polymers which degrade at low temperatures in the melting range. Polyvinylchloride belongs to this group. The base-line definition is always a problem due to early thermal degradation and the low overall crystallinity which leads to melting over a very broad temperature range.

The kinetics of crystallization, which is of interest for both academic and industrial reasons, is preferably studied under isothermal conditions by DSC dilatometry or TOA. These methods reveal the overall crystallinity, volume(v,) or mass(w,) crystallinity as a function of time(t), and the general Avrami equation can be applied:

$$1-\frac{v_c}{v_\infty}=e^{-Kt^n} \tag{23.1}$$

where n and K are constants and ν_∞ is the maximum crystallinity attained.

The Avrami exponent (n) depends on nucleation type, the geometry of crystal growth and the kinetics of crystal growth. The kinetics at low degrees of conversion usually follows the Avrami equation but deviates from the linear trend in the plot. Most of the kinetic work has dealt with the temperature dependence of the growth rate in accordance with the kinetic theory of Lauritzen and Hoffman(see Hoffman et al. 1975). The experimental data, the linear growth rate (G) of spherulites (axialites), are obtained by hot-stage polarized light microscopy at different constant temperatures.

1. New words

lamella[lə'melə]*n.* 薄板,薄片,薄层

superheating[ˌsjuːpə'hiːtiŋ] *n.* [热] 过热,市场狂热;*v.* 过度加热

immiscible[i'misib(ə)l] *adj.* 不融和的,不能混合的

rigorous['rig(ə) rəs] *adj.* 严格的,严厉的,严密的

chlorinated['klɔrinetid] *adj.* 含氯的

spherulite['sferjulait] *n.* [地质]球粒

bimodal[bai'məud(ə)l] *adj.* 双峰的

polymorphs['pɔlimɔːfs] *n.* [晶体]多形体

2. Phrases and expressions

the order of 顺序

as a function of time 作为时间的函数

zero-entropyproduction 零熵产生

in accordance 与……一致

3. Notes to the text

① The crystal thickening of polymer crystals expected to occur on the basis of these thermodynamic arguments has been verified by X-ray diffraction experiments. 通过 X 射线衍射实验验证了这些热力学论证对聚合物晶体增厚的预测。

② These micrometre-thick crystals display a distinctly different melting behaviour from that of the thin folded-chain single crystals grown from solution. 这些微米厚的晶体显示出明显不同于从溶液中生长出来的薄的折叠链单晶的熔化行为。

③ Samples which undergo these premelting transitional phenomena display bimodal or multimodal melting provided that the original crystals are of about the same lamella thickness. 在原晶片厚度基本相同的条件下,试样在预熔过渡过程中出现双峰或多峰熔化现象。

④ The melting of the once-folded orthorhombic crystals of n-$C_{294}H_{590}$ leads to recrystallization into extended-chain crystals which melt at a higher temperature in the high-temperature melting peak. 一次折叠正交晶 n-$C_{294}H_{590}$ 的熔融导致再结晶为长链晶体,在高温熔融峰温度较高时熔化。

4. Exercises

(1) Explain thermal behavior of semicrystalline polymers.

(2) Demonstrate crystallization kinetics of a semicrystalline polymer using the general Avrami equation.

Reading Material

Thermal Behavior of Amorphous Polymers

The glass transition temperature is possibly the most prominent temperature for an amorphous polymer. The stiffness of a typical amorphous polymer changes by three orders of magnitude from a few gigapascals in the glassy state (low-temperature side) to a few megapascals in the rubbery state (high-temperature side)①. The glass transition appears at first sight to be a second-order phase transition, i. e. volume (V) and enthalpy (H) are continuous functions through the transition temperature interval (Figure 23. 5).

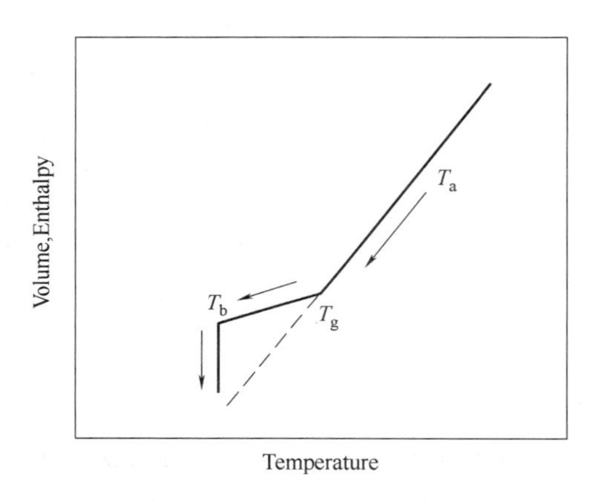

Figure 23. 5 Schematic representation of the glass transition in amorphous polymers. The decrease in volume and enthalpy at T_b is referred to as physical ageing

However, if the amorphous polymer is cooled from temperature T_a, a break in the rectilinear H, $V = f(T)$ curve appears at the glass transition temperature (T_g). If the cooling is stopped at a temperature near T_g, at T_b in Figure 23. 5, and the sample is kept at this temperature, the enthalpy and volume of the sample decrease as a function of time. This so-called physical ageing clearly shows that glassy amorphous polymers are not in equilibrium and that the measured glass transition temperature is a 'kinetic' temperature rather than a temperature associated with a true thermodynamic transition. Physical ageing thus leads to more densely packed material of higher stiffness and lower impact strength. The kinetics of physical ageing can be followed by DSC/DTA and dilatometry. The fundamental aspects and practical implications of physical ageing have been the subject of many papers. Physical ageing was reviewed by Struik (1978).

The non-equilibrium phenomenon described above also leads to a significant dependence on cooling rate of the recorded T_g. Atactic polystyrene, for example, shows a change in T_g from 365K at a cooling rate of 1K/h to 380K at a cooling rate of 1K/s. Even with a cooling rate of 1K/s the observed T_g does not decrease below 351K. It is important to note that these data refer to the T_g recorded on cooling. On heating, glassy amorphous polymers exhibit superheating effects. If a slowly cooled amorphous polymer is rapidly heated, the T_g is shifted to significantly higher temperatures and an endothermic hysteresis peak is observed above the glass transition. In conclusion, the most reliable and precise way of measuring T_g is by cooling the melt at a specified low cooling rate and recording the step in the specific heat curve.

The morphology of amorphous polymer blends is indeed of great practical importance. The impact strength of stiff and brittle glassy polymers (e. g. atactic polystyrene) may be greatly improved by including a few per cent of an immiscible elastomer such as polybutadiene②. Provided that the polymers have different glass transition temperatures, DSC (DTA) or TMA are valuable tools for seeking the answer to the crucial question of whether the polymers are miscible or whether they exist in different phases.

Figure 23. 6 shows the thermogram obtained from an immiscible mixture of styrene-acrylonitrile (SAN) copolymer and polybutadiene. This blend is a high impact strength material referred to as acrylonitrilebutadiene-styrene (ABS) plastic. Two glass transitions are observed, the low-temperature transition associated with the polybutadiene and the hightemperature transition associated with the SAN copolymer. Miscible blends are less commonly found. One of the most studied blends is that between polystyrene and poly (phenylene oxide) (PPO). This blend is miscible in all proportions and the films of the blends are optically clear. Only one glass transition intermediate in temperature between the $T_g's$ of polystyrene and PPO has been reported. The compositional dependence of the T_g of a compatible binary blend follows in some cases, e. g. the Fox equation for blends of polyvinylchloride and ethylene-vinyl acetate copolymers:

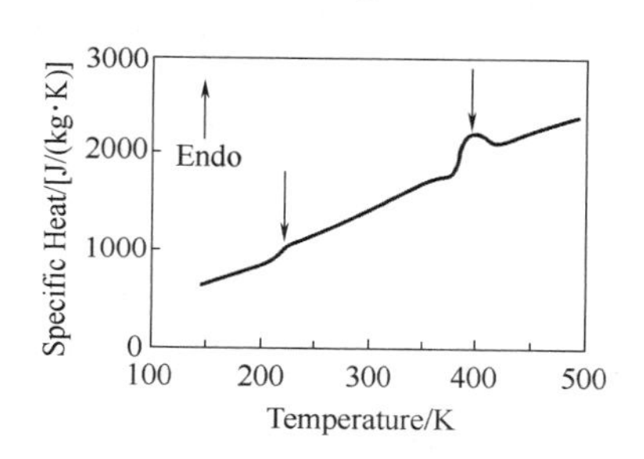

Figure 23. 6 DSC thermogram of ABS showing two glass transitions. Drawn after data from Bair (1970)

$$\frac{1}{T_g} = \frac{\omega_1}{T_{g1}} + \frac{\omega_2}{T_{g2}} \tag{23.2}$$

where ω_i and T_{gi}; are the weight fractions and glass transition temperatures of the pure polymers. Several other equations relating T_g to the composition have also been proposed.

1. New words

gigapascal [gigə 'pæsk(ə)l] *n.* 吉帕斯卡(物理单位)

dilatometry [dilə'tɔmitri] *n.* [分化][物]膨胀测定法,膨胀法

atactic [e'tæktik] *adj.* 不规则的,[有化]无规立构的

endothermic [ˌendəu'θɜːmik] *adj.* [热]吸热的,温血的

hysteresis[ˌhistəˈriːsis] *n.* 迟滞现象，滞后作用，磁滞现象

2. Phrases and expressions

second-order phase transition 二级相变点
physical ageing 物理老化
styrene-acrylonitrile 苯乙烯丙烯腈
ethylene-vinyl acetate 乙烯-醋酸乙烯酯

3. Notes to the text

① The stiffness of a typical amorphous polymer changes by three orders of magnitude from a few gigapascals in the glassy state(low-temperature side) to a few megapascals in the rubbery state(high-temperature side). 典型的非晶聚合物的硬度变化三个数量级，从玻璃态(低温侧)的几十亿帕斯卡到橡胶态(高温侧)的几百万帕斯卡。

② The impact strength of stiff and brittle glassy polymers(e. g. atactic polystyrene) may be greatly improved by including a few per cent of an immiscible elastomer such as polybutadiene. 刚性和脆性玻璃聚合物(如无规聚苯乙烯)的冲击强度可以通过加入少量不相容弹性体(如聚丁二烯)而大大提高。

Lesson 24 Thermal Behavior of Liquid-Crystalline Polymers

Liquid-crystalline polymers are a relatively new group of polymers which have aroused considerable interest during the last decade. Main-chain polymers, with the mesogenic stiff units implemented in the main chain, exhibit a unique combination of good processing and good mechanical properties and hence are now used as engineering plastics. Side-chain polymers, with the mesogenic in the side chains, are mainly used in speciality polymers with potential application in electronics and optronics①. Liquid-crystalline polymers exhibit a number of thermal transitions, as shown in Figure 24. 1.

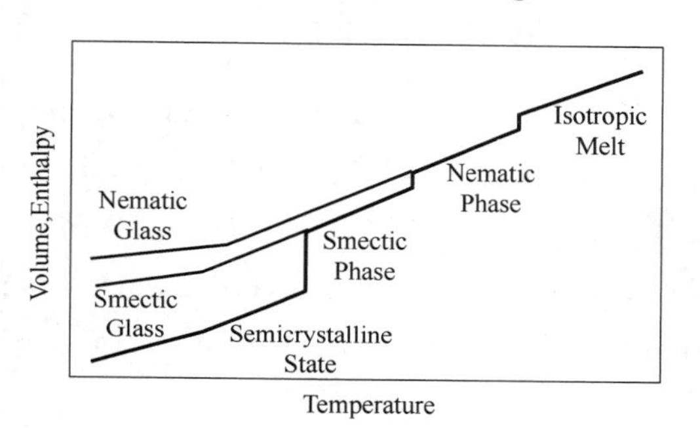

Figure 24. 1 Typical phase transitions in thermotropic liquid-crystalline polymers

Thermal analysis, DSC/DTA, TMA and TOA and hot-stage microscopy are, together with X-ray diffraction, the popular methods for structural assessment at different temperatures. A combination of these methods is commonly used. X-ray diffraction of aligned samples is the most reliable method for structural assessment. When the isotropic melt cools from very high temperatures, it is ultimately transformed into a liquid-crystalline phase, i. e. a so-called mesophase. A great number of different liquid-crystalline structures have been reported. Two major groups, differing in degree of order, exist: nematics and smectics. The high-temperature transition is readily revealed by DSC(DTA) as an exothermic first-order transition (see Figure 24. 1) and by TOA and hot-stage polarized microscopy from the formation of birefringent structures. The enthalpy involved in an isotropic-nematic transition is significantly smaller than that involved in an isotropic-smectic transition②. The phase assignment of the mesophase can be achieved by polarized light microscopy according to the scheme by Demus and Richter (1978) or preferably by X-ray diffraction.

At lower temperatures, a number of liquid crystalline transitions may occur which again can be recorded by DSC/DTA as exothermic first-order transitions (Figure 24. 1). Hot-stage microscopy and X-ray diffraction are used to determine the nature of these transitions. At lower temperatures, solid crystals may be formed. The latter are revealed by DSC/DTA as an exothermic first-order transition, by TMA as an increase in sample stiffness and by X-ray diffraction as sharp Bragg reflections. Some liquid-crystalline polymers, e. g. copolyesters, are supercooled to a glassy state without crystallizing.

1. New words

mesogenic[mesəu'dʒenik] *adj.* 液晶的

optronics[ɔp'trɔniks] *n.* 光电子学，光导发光学，光电产品

mesophase ['mesəufeiz] *n.* 中间相，液晶相

nematic[ni'mætik] *adj.*（液晶）[晶体] 向列的

smectic['smektik] *adj.* 近晶的，净化的

birefringent[ˌbairi'frindʒənt] *adj.* [光] 双折射的

2. Phrases and expressions

liquid-crystalline 液晶

bragg reflections 布拉格反射

3. Notes to the text

① Side-chain polymers, with the mesogens in the side chains, are mainly used in speciality polymers with potential application in electronics and optronics. 侧链聚合物主要用于具有电子和光电子潜在应用前景的特种高分子材料中，侧链中含有液晶体。

② The enthalpy involved in an isotropic-nematic transition is significantly smaller than that involved in an isotropic-smectic transition. 各向同性向列的转变的焓比各向同性近晶的转变的焓要小得多。

4. Exercises

(1) Demonstrate the thermal behavior of liquid-crystalline polymers.

(2) Explain typical phase transitions in thermotropic liquid-crystalline polymers.

Reading Material

Polymer Degradation

Polymers are, with few exceptions, very sensitive to degradation reactions occurring both during the melt-processing and during use. Reactions with oxygen, thermal oxidation and photo-oxidation, are for many polymers the dominating degradation reactions. Stabilizers, e. g. antioxidants, increase the stability of polymers and extend the life of many products considerably. The exposure of a plastic material to heat and oxygen leads to consumption of the antioxidant. It has therefore been important to develop efficient methods for the determination of antioxidant content. Thermal analyses, DSC/DTA and TG, are among the most frequently used methods for this purpose.

The oxidative induction time (OIT) is measured by first heating the polymer sample while keeping it in a nitrogen atmosphere to a high temperature, typically 200℃ for polyethylene①. After the establishment of constant temperature, the atmosphere is switched to oxygen and the time (OIT) to the start of an exothermic (oxidation) process is measured (Figure 24.2). It has been shown that the OIT exhibits an Arrhenius temperature dependence and that there is a linear relationship between OIT and the content of efficient antioxidant. This follows from the fact that the consumption of antioxidant at these conditions follow zero-order kinetics.

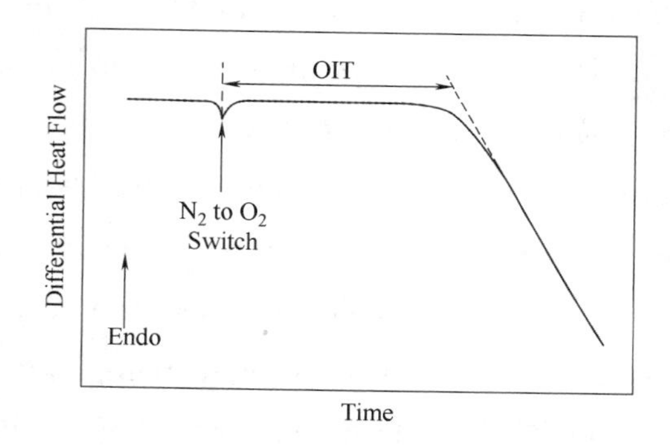

Figure 24.2 Typical thermogram from an OIT measurement

It is important to note that, in order to use OIT data as an absolute method for the determination of the antioxidant content, the system must be calibrated to determine the constants k and activation energy. Similar measurements can be made by TG. The first indication of oxidation is a small increase in sample mass. At a later stage, when degradation leads to the formation of volatile products, the sample mass decreases strongly.

When the constants k and activation energy have been established, the antioxidant content can be indirectly measured by recording the thermogram at a constant heating rate in oxygen atmosphere and determining the oxidation temperature (T_{ox}) as the temperature of the exothermic deviation from the scanning base line[②].

An example is presented in Figure 24.3. Since OIT is proportional to C_0, it is evident that the dynamic method is less suitable for determination of antioxidant content in highly stabilized systems (high C_0), but it is sufficiently sensitive for small variations in C_0 in systems with small C_0 values.

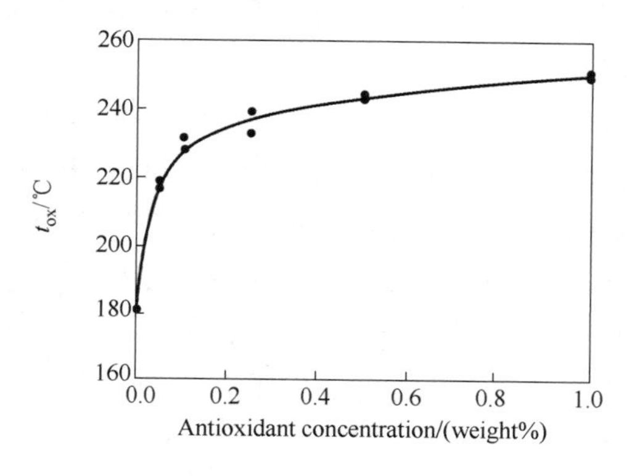

Figure 24.3 Oxidation temperature as a function of antioxidant concentration (lrganox 1010 in a medium density polyethylene). With permission from Elsevier (Karlsson, Assargren and Gedde 1990)

OIT/ T_{ox} measurements provide rapid and reliable results. Many materials contain a system of antioxidants, often a combination of a primary antioxidant (hindered phenol or amine) and a secondary antioxidant (phosphite, thioester, etc.). Thermal analysis provides no information about the concentration of the different antioxidants separately but rather an overall assessment of the stability. Extraction followed by chromatography (HPLC) is one of the main techniques for the determination of antioxidant concentration, but it is no doubt much more time-consuming than DSC/DTA.

Thermal analysis involves a set of analytical methods by which a physical property of a sample is measured as a function of temperature (time). The properties that are studied are typically enthalpy (DSC/DTA), dimensions (TMA, dilatometry), visco-elastic properties (DMTA), mass (TG), dielectric properties (DETA) and optical properties (TOA). Reproducible and accurate results are currently available and allow a great number of materials and phenomena involving both physical and chemical aspects to be studied. Thermal transitions of polymers, involving both crystallization/melting and glass formation, are irreversible processes and great care must be taken in the interpretation of the data.

1. New words

stabilizer ['steibilaizə] *n.* [助剂]稳定剂,稳定器

antioxidant [ænti'ɔksid(ə)nt] *n.* [助剂]抗氧化剂

phosphate ['fɔsfeit] *n.* [无化]亚磷酸盐

thioester [ˌθaiəu'estə] *n.* 硫酯,[有化]硫代酸酯

chromatography [ˌkrəumə'tɔgrəfi] *n.* 色层分析,色谱分析法

2. Phrases and expressions

proportional to　与……成比例

irreversible process　不可逆过程

3. Notes to the text

① The oxidative induction time (OIT) is measured by first heating the polymer sample while keeping it in a nitrogen atmosphere to a high temperature, typically 200℃ for polyethylene. 氧化诱导时间(OIT)是衡量保持它在氮气氛中第一次加热聚合物样品到高温,通常加热聚乙烯200℃。

② When the constants k and activation energy have been established, the antioxidant content can be indirectly measured by recording the thermogram at a constant heating rate in oxygen atmosphere and determining the oxidation temperature (T_{ox}) as the temperature of the exothermic deviation from the scanning base line. 当常数 k 和活化能确定后,通过记录氧气氛中恒定升温速率下的热像图,将氧化温度(T_{ox})确定为与扫描基线放热偏差的温度,间接测定抗氧化剂含量。

PART 6 POLYMER MATERIALS

Lesson 25 Plastics

Plastics are synthetic or man-made materials which have unique and remarkable properties. Many different plastics may be obtained by manipulating the molecules and changing the chemical combinations.

Plastics have important roles in agriculture, appliances, clothing, construction, electronics, furniture, packaging, transportation, and numerous other areas. It is hardly possible to talk about the future for plastics except in highly optimistic terms.

Plastics is a multibillion-dollar industry which produces synthetic materials and products, many of which were never dreamed of only a few years ago. Many different plastics may be obtained by manipulating the molecules and changing the chemical combinations. Today we would be utterly lost without the synthetic materials (artificial resins produced by chemical reaction of organic substances). Many products are made of plastics produced at less cost than was possible with natural materials obtained from the earth.

Plastics have become a part of society which accepts and often takes for granted the major role of these fantastic materials[①] (Figure 25.1). Plastics have important roles in agriculture, appliances, clothing, construction, electronics, furniture, packaging, transportation, and numerous other areas. The automobile contains hundreds of plastic parts (Figure 25.2 and Figure 25.3). The home, school, office and industry, rely heavily upon plastic parts for television sets, computer, telephones and furniture; also plastic products such as carpeting and containers. Even the football player relies on plastics. If all of the plastics in his uniform were removed, he would have little left for

Figure 25.1 Moon walking requires shoes that will withstand freezing cold and scorching heat. These flexible plastic silicone shoes provide excellent insulation and withstand moon temperatures without melting or becoming brittle

protection.

This is truly the age of plastics. In the future plastics will make possible things that have never been done before. New designs, new shapes, new applications stemming from plastics' astonishing and limitless versatility will permit beauty of line and purpose that is sure to become the hallmark of a new American culture②.

Figure 25.2 This automobile interior features a plastic steering wheel, dash panel, upholstery and control console

Figure 25.3 Plastics are used for many exterior parts of automobiles. This picture illustrates the use of plastics for the entire body and trim

It might appear that the plastics industry is a manufacturing group unrelated to many other industrial organizations. However, the plastics industry relies heavily upon many other industries and is an integral part of numerous related industries. Many products of other materials involve the use of plastics. For example, metal products are often plastic coated. Pieces of wood are fastened together with plastic adhesives and coated with plastic finishes. Baseball and football fields are being covered with plastic imitation grass.

There exists in the field of plastics, a tremendous shortage of skilled employees and a real need for an expanded program in plastics education at all levels.

1. New words

combination[kɔmbi'neiʃən] *n.* 组成
artificial[a:ti'fiʃəl] *adj.* 人造的，模拟的
fantastic[fæn'tæstik] *adj.* 奇异的，幻想的
astonish[əs'tɔniʃ] *v.* 使惊讶，惊奇
scorching['skɔ:tʃiŋ] *adj.* 灼热的，激烈的
insulation[ˌinsju'leiʃən] *n.* 绝缘
dash[dæʃ] *n.* 仪表板(= dash panel)
console[kən'səul] *n.* 控制台
trim[trim] *adj.* 整齐的，整洁的； *vt.* 整理，修整，装饰
versatility[və:sə'tiliti] *n.* 通用性
hallmark['hɔ:l'ma:k] *n.* 标记
manipulation[məˌnipju'leiʃən] *vt.* 处理，操作，控制
stem[stem] *vi.* 起源，发生

2. Phrases and expressions

dreamed of 梦想
chemical combination 化学组成

organic substance 有机物
dash panel （汽车的）仪表板
control console 控制［操纵］台
take sth. for granted 认为……是理所当然
it might appear 看来，似乎
integral part 组成部分
stem from 由……发生（产生，引起）
rely upon（rely on） 依靠

3. Notes to the text

① Plastics have become a part of society which accepts and often takes for granted the major role of these fantastic materials. 塑料这种奇异材料所起到的重要作用，必然为人类所接受并成为社会的一部分。

② New designs, new shapes, new applications stemming from plastics' astonishing and limitless versatility will permit beauty of line and purpose that is sure to become the hallmark of a new American culture. 由于塑料具有极其惊人和极广泛的通用性，由此产生的许多新设计、新款式和新应用将会使各种产品及其应用变得更加完美，从而可以确信，它必将成为美国新文化的标志。

4. Exercises

(1) What kind of industry is plastics?

(2) Make a list of the newest plastics products which are available in your home. Explain what properties of the material seem outstanding for each product.

(3) Make a collection of all the standard shapes of plastics products you can find such as sheets, rods, tubes, etc.

(4) Examine an automobile of the latest model and make a list of the parts you find that are made of plastics. Explain why you think plastics are used for each part.

Reading Material

The Historical Development of Plastics Industry

The plastics industry is one of the newest of the major industries. Since plastics are synthetic, or man-made materials, it is easy to understand why the industry followed other material industries in development. Man had depended upon wood, metals, concrete, glass and other natural materials for centuries.

It was not until 1868 that the first plastic material was commercially produced. The need for replacing ivory for billiard balls led John Wesley Hyatt, a printer, to experiment with a new process, the reaction of camphor on cellulose nitrate. The result was a material that could be formed in sheets but was not suitable for molding. Called cellulose nitrate, this plastic later became known as "Celluloid." Celluloid was quickly adopted for many purposes. It was used for windows in early automobiles and became widely used for motion picture film. Cellulose nitrate, however, was highly flammable and was later replaced by plastics which would not easily burn.

It was not until 1909 that another synthetic material appeared commercially. In that year, Dr. Leo Baekeland announced a new resin, phenol formaldehyde, which was to become a major plastic in industry. Given the name Bakelite, this new plastics material could be molded using heat and pressure, to form high heat resistant products such as coffee-pot handles (Figure 25.4), and electrical outlet plugs.

Figure 25.4 The heat resistant parts for this electric coffee maker were made from phenolic molding compounds

This was the beginning of a rapidly developing science of synthetic materials. As the years passed, new techniques along with new scientific discoveries enabled chemists to introduce new plastics with ever increasing properties. In 1927 cellulose acetate was produced. Injection molding added a new dimension to production of items from this material①. Rapid development of vinyl resins followed by polystyrene and polyethylene led to an overwhelming volume of new plastics being introduced. Table 25.1 lists development dates of commercially important plastics of the industry. Every year new plastic materials are being discovered and created and there is continued innovation in existing plastics. Plastics are giving man an opportunity to meet his environmental needs precisely.

Table 25.1 Development of Plastics materials

Date	Material	Typical products
1868	Cellulose Nitrate	Guitar Picks
1909	Phenol-Formaldehyde	Telephone Handsets, Distributor Caps
1927	Cellulose Acetate	Blister Packages
1927	Polyvinyl Chloride	Shower Curtains
1929	Urea-Formaldehyde	Lighting Fixtures
1935	Ethyl Cellulose	Cosmetic packages
1936	Acrylic	Lighting Displays
1936	Polyvinyl Acetate	Adhesives
1938	Cellulose Acetate Butyrate	Screwdriver Handles
1938	Polystyrene	Refrigerator Parts
1938	Nylon	Gears, Brush Bustles
1939	Polyvinylidene Chloride	Wood Wrap
1939	Melamine-Formaldehyde	Dinnerware
1941	Alkyd	Electron Tube Bases
1942	Polyester	Boat Hulls

Continuation

Date	Material	Typical products
1942	Polyethylene	Squeezable Bottles
1943	Fluorocarbon	Metal Coatings, Bearings
1943	Silicone	Gaskets and Sealants
1945	Cellulose Propionate	Ball Point Pens
1947	Epoxy	Molds and Tools
1948	Acrylonitrile-butadiene-Styrene	Pipe and Pipe fitting
1949	diallyl Phthalate	Circuit Breakers
1954	Polyurethane	Foam Cushions
1956	Acetal	Tool Handles
1957	Polypropylene	Television Cabinets
1959	Polycarbonate	Street Light Globes
1962	Phenoxy	Bottles
1962	Polyallomer	Luggage
1964	Ionomer	Skin Packages, Safety Glasses
1964	Polyphenylene Oxide	Surgical Tools
1964	polyamide	Bearings
1965	Parylene	Protective Coatings
1965	Polysulfone	Electronic Parts

1. New words

ivory['ɑivəri] *n.* 象牙
billiard['bilɪrd] *adj.* 台球的
camphor['kæmfə] *n.* 樟脑
nitrate['naitreit] *n.* 硝酸盐
celluloid['seljulɔid] *n.* 赛璐珞
phenol['fi:nɔl] *n.* 苯酚
phenolic[fi'nɔlik] *n.* 酚醛塑料，酚醛树脂，酚类(的)
formaldehyde[fɔ:'mældihaid] *n.* 甲醛
bakelite['beikəlait] *n.* 电木，胶木
acetate['æsitit] *n.* 醋酸盐
overwhelming[ˌəuvə'hwelmiŋ] *adj.* 优势的
innovation[inəu'veiʃən] *n.* 创新，改革
butyrate['bju:tireit] *n.* 丁基盐(酯)
alkyd['ælkid] *n.* 醇酸
fluorocarbon[ˌflu:ərəu'ka:bən] *n.* 碳氟化合物
phthalate[θæleit] *n.* 酞酸盐(酯)
acetal['æsitæl] *n.* 聚甲醛，缩醛
phenoxy['finɔksi] *n.* 苯氧基
polyallomer['pɔli'ælɔmə] *n.* 同质异晶聚合物
ionomer[ˌaiə'nəmə] *n.* 含离子键的聚合物
parylene['pa:rɪli:n] *n.* 聚对二甲苯
polysulfone[ˌpɔli'sʌlfəun] *n.* 聚砜

2. Phrases and expressions

cellulose nitrate　硝酸纤维素
phenol-formaldehyde　苯酚-甲醛
science of synthetic　合成材料学科
cellulose acetate　醋酸纤维素
vinyl resin　乙烯基树脂
melamine-formaldehyde　三聚氰胺-甲醛树脂
polyphenylene oxide　聚氧化乙烯

3. Notes to the text

① Injection molding added a new dimension to production of item from this material. 注塑模塑使利用这种材料生产产品又增加了新的方法。

Lesson 26 Thermoplastic Materials and Polyethylene Resins

Thermoplastic Materials:

It is important to become familiar with the outstanding physical and chemical properties of thermoplastic plastics so you can make wise selections for their use. You should understand how and why these materials can be processed in certain ways. The relationship between the properties of each polymer and how these properties affect the way in which they can be molded, and why they are selected for certain products, is a key to understanding the plastics industry[①]. Always keep in mind the following three factors as you study plastics materials:

a. The outstanding properties of the resin.

b. How these properties determine the ways in which it can be molded.

c. Properties which make the resin suitable for specific products.

Polyethylene:

Polyethylene is the major member of a group of chemical compounds known as polyolefins. It is one of the most widely used polymers of any of the thermoplastic materials. To make polyethylene, it is necessary to use high-purity ethylene gas. The ethylene gas can be made from natural gas or obtained as a by-product of a petroleum refinery. Through addition polymerization, the resulting polymer has the following basic structure.

$$\left[CH_2—CH_2 \right]_n$$

Polyethylene is produced in two forms in terms of density. Low and intermediate density polyethylenes are produced with a relatively short chain molecular structure and a high degree of side branching. This structure provides a polymer with approximately 65 percent crystallinity. On the other hand, high-density polyethylene is polymerized to form much longer linear chains with few side branches. This results in a greater density and a crystallinity range of 85 percent. Polyethylene can best be described as a flexible, tough, chemically resistant, crystalline polymer (Figure 26.1).

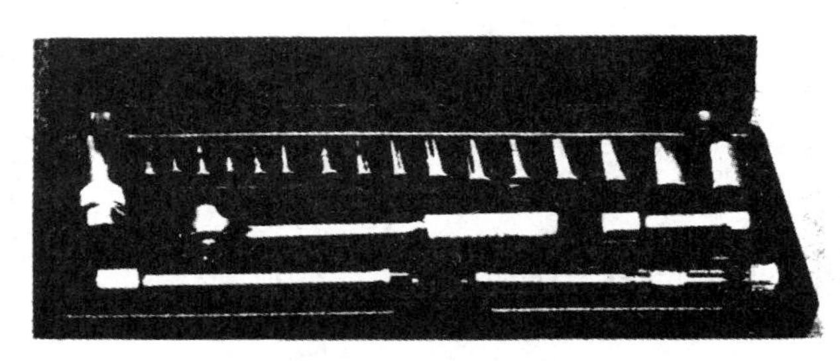

Figure 26.1 The properties of high density polyethylene meet the design requirements of this rugged tool case. The case contains a built-in hinge and snap clasp

Polyethylene appears in its

natural form as a milky white, waxy feeling material. In general, as the density increases, the stiffness, hardness, strength, heat distortion point, and ability to transmit gasses increases. As density decreases, impact strength and stress crack resistance increases. Stress cracking is a surface change that polyethylene, and some other plastics, undergo when exposed to oils, gasoline and other hydrocarbons② (Figure 26.2). It appears on the material as a flaky, cracked surface.

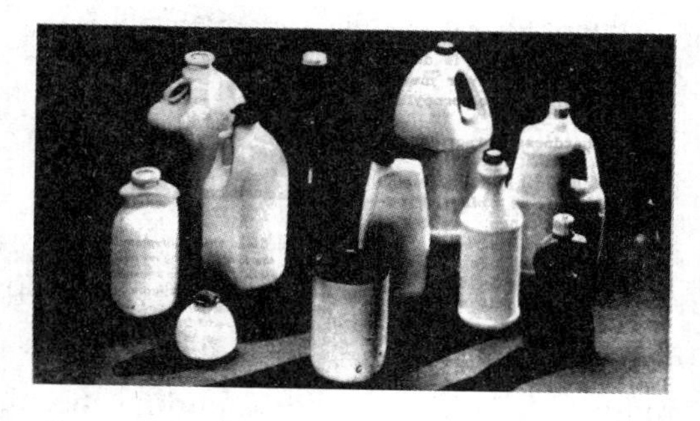

Figure 26.2 High density polyethylene is used extensively for bottles and containers because of its ease of processing, toughness, and economy

Polyethylene not only finds many product applications due to its properties but also because of the many forms in which it is produced as a resin. It may be obtained in granules, powders, film, rod, tube and sheet form and molded through such processes as injection molding, fluidized bed coating, blow molding, extrusion, vacuum forming, casting and calendering. Polyethylene is used for many purposes, including: containers, electrical insulation, housewares, chemical tubing, toys, freezer bags, flexible ice cube trays, snap-on lids and battery parts. Two major applications of polyethylene are films for packaging soft goods and other non-perishables and blow molded bottles. Squeeze bottles from low-density polyethylene and detergent bottles from high density are typical products.

1. New words

polyolefin['pɔli'əuləfin] *n.* 聚烯烃
refinery[ri'fainəri] *n.* 精炼厂
density['densiti] *n.* 密度
rugged['rʌgid] *adj.* 高低不平的，崎岖的，粗糙的，有皱纹的
tough[tʌf] *adj.* 韧性的，刚性的
built-in[ˌbilt'in] *adj.* 内置的，固定的，嵌入的； *n.* 内置
hardness['ha:dnis] *n.* 硬度
distortion[dis'tɔ:ʃən] *n.* 变形，畸变，挠曲
tray *n.* 托盘
fluidize['fluidaiz] *vt.* 流化
casting['ka:stiŋ] *n.* 铸塑，流涎
calender['kælində] *v.* 压延
houseware['hauswεə] *n.* 家庭用具
squeeze[skwi:z] *v.*; *n.* 挤压
nonperishable[nɔn'periʃəbl] *adj.* 不易腐烂的
detergent[di'tə:dʒənt] *adj.* 清洁的；*n.* 清洁剂

2. Phrases and expressions

keep in mind 记住，放在心里
side branch 支链，分枝
impact strength 冲击强度
stress crack 应力开裂

fluidized bed coating 流化床涂布
blow molding 吹塑模塑
vacuum forming 真空成型
chemical tubing 化工管道
ice cube (加入饮料用的)小方冰块
snap-on lid 带按扣的盖子

3. Notes to the text

① The relationship between the properties of each polymer and how these properties affect the way in which they can be molded, and why they are selected for certain products, is a key to understanding the plastics industry. 了解塑料工业的关键是了解每种高聚物的性能和这些性能怎样影响其模塑方法之间的关系,以及了解为什么要针对一定的产品来选择高聚物。

② Stress cracking is a surface change that polyethylene, and some other plastics, undergo when exposed to oils, gasoline and other hydrocarbons. 应力开裂现象是指聚乙烯或一些其他的塑料与油、汽油及其他烃类物质接触时所产生的表面变化情况。

4. Exercises

(1) Translate the text into Chinese.

(2) How can polyethylene best be described as to properties?

(3) What is meant by stress cracking?

(4) Does polyethylene film transmit gasses or does it resist them?

(5) Prepare a chart showing the different ways in which polyethylene is used for products according to the properties it displays.

(6) Visit a local plastics plant (if available) and write a report on the products they manufacture, the various plastics they use, and the processes they are using to produce such products.

(7) Translate the following sentences into Chinese and note which method is converted.

a. The mould is a well-designed structure.

b. The molting point of polymer is obviously a function of polymer structure.

c. There can be greater difficulties with changes in the structure of the polymer.

d. The polymers are blended for (the purpose of) improving their mechanical properties.

Reading Material

Polypropylene Resins

Polypropylene resins are made from propylene gas addition polymerization. The process is similar to the production of the high density polyethylene. This yields a high molecular weight and high crystalline polymer with no side branches. The newer polypropylenes are isotactic in structure with a very high crystallinity compared to the former amorphous polymers. The long chain molecule of polypropylene with large side groups is represented in its linear form.

$$\left[CH_2 - \underset{\displaystyle CH_3}{\underset{|}{CH}} \right]_n$$

Polypropylene is the second resin in the family of polyolefins composed of long chain saturated hydrocarbons. The readily available source of propylene gas from the petroleum industry along with the improved properties over polyethylene make it one of the leading polymers on the industrial market①.

Polypropylene is produced as a molding material, as granules. In its natural state it is a fairly hard, cloudy white material. Because of the high degree of crystallinity in the polymer, it cannot be produced as a crystal clear material. However, when it is produced as film its clarity is similar to polyethylene film. It is one of the lightest plastics available with a density range of 0.890 to 0.905. This is often considered an advantage since, like polyethylene, it will float.

The improved properties of polypropylene over the same properties of polyethylene are②: rigidity, flex life, heat distortion, tensile strength and stress crack resistance. The increase in these properties is due to higher crystallinity and the larger methyl group. The major properties of polypropylene are:

(1) Good surface hardness and scratch resistance.

(2) Excellent dimensional stability.

(3) Outstanding flex life as hinge. Products with integral hinges are one of its most valued uses (Figure26.3).

(4) Excellent electrical properties even at high heat.

(5) Some hydrocarbons will soften and swell the polymer.

(6) Tough at temperatures from 105 ℉ to 15 ℉ but brittle below 0 ℉.

(7) Excellent resistance to water and gas vapor.

Figure 26.3 This self-hind polypropylene painting kit weighs less than two pounds. It is tough, durable, and waterproof

(8) Good chemical resistance.

(9) Serviceable above sterilization temperatures, 212 ℉.

(10) Easily colored in opaque and translucent products.

The combination of properties of polypropylene makes it an exciting material for the product designer. It is suitable for a wide variety of molding processes. It can be injection molded, blow molded, thermoformed, extruded into sheet, film, pipe, wire coatings and fibers and processed by a number of other techniques. Polypropylene finds use in applications such as luggage cases, card files, cosmetic cases, and automobile accelerator pedals, which take advantage of its flex life and impact strength. It is used in hospital equipment because it is sterilizable, resistant to chemicals, and transparent in thin sections.

Figure 26.4 Glass filled polypropylene is used for this pump housing, the impeller, magnet housing, and exhaust chamber. Polypropylene was chosen because of its chemical resistance, ability to allow magnetic force to pass through it (not possible with some metals), and its self-sealing for assembly

Polypropylene has also found use in products such as blown bottles, fibers for carpeting, housewares, electronic parts, aviation components, and the packaging industry (Figure 26.4) illustrates the use of polypropylene in a pump housing.

These product properties will become more apparent to you as you process and test sheet material or produce molded parts.

1. New words

propylene['prəupili:n] *n.* 丙烯
clarity['klæriti] *n.* 透明，清晰度
float[fləut] *v.* 浮起
rigidity[ri'dʒiditi] *n.* 刚性，刚度
flex[fleks] *n.* 挠曲
tensile['tensail] *adj.* 可拉伸的
scratch[skrætʃ] *n.* 刮伤
durable['djuərəbl] *adj.* 持久的，耐用的
methyl['meθil] *n.* 甲基
thermoform['θə:məufɔm] *n.* 热成型；*vt.* 使加热成型，给……用热力塑型
accelerator[æk'seləreitə] *n.* 加速器
pedal['pedl] *n.* 踏板
sterilizable[ˌsterə'lizəbl] *adj.* 杀菌的
section['sekʃən] *n.* 型材

2. Phrases and expressions

crystalline polymer 结晶聚合物
composed of 由……组成的
saturated hydrocarbon 饱和烃
natural state 自然状态
crystal clear material 晶状透明材料
flex life 挠曲寿命
methyl group 甲基
scratch resistance 抗刮伤
find use in 应用于
cosmetic cases 化妆盒
aviation component 航空构件
sheet material 片材
molded part 模塑制件

3. Notes to the text

① The readily available source of propylene gas from the petroleum industry along with the improved properties over polyethylene make it one of the leading polymers on the industrial market. 丙烯气体在石油工业中的稳定来源以及比聚乙烯更好的性能使其在工业市场中成为一种主要的高聚物。

② The improved properties of polypropylene over the same properties of polyethylene are:…. 与聚乙烯相比,聚丙烯所改善的性能是……。

Lesson 27 Vinyl Resins and Polyvinyl Chloride

Vinyl resins cover a broad group of materials that range in properties from hard rigid products to soft flexible formulations. They are derived from the "vinyl" radical: $CH_2{=}CH_2$. Various atoms or side groups may be attached to the vinyl radical to produce polymers of varying properties.

The largest polymer in the vinyl group is polyvinyl chloride. Its structure is represented by this chemical diagram:

$$\left[CH_2 - \underset{\displaystyle Cl}{\underset{|}{CH}} \right]_n$$

PVC is produced commercially from acetylene and hydrogen chloride and can be compounded to give almost any degree of flexibility to the final product by adding[①] plasticizers, fillers, and stabilizers. The structure of PVC is only slightly crystalline due to its atactic molecular arrangement which provides good clarity to the material.

The general characteristics of all vinyls are similar, such as good strength, excellent water and chemical resistance and unlimited color possibilities. PVC exhibits self-extinguishing characteristics, good weather resistance, electrical properties and abrasion resistance. Two factors that make PVC uniquely different from other plastics are its wide range of properties and being self-extinguishing. The compounder can produce materials that are hard and rigid to those that are soft and flexible. This explains why polyvinyl chloride is used in so many product applications.

Two examples of products taking advantage of the properties of PVC are flexible sheet material and rigid pipe and tubing. Calendered sheet is used for simulated leather

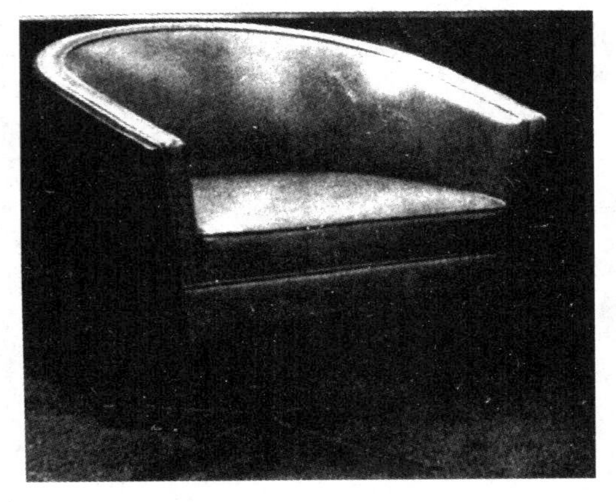

Figure 27.1 Office equipment makes use of flexible polyvinyl chloride simulated leather upholstery for chair coverings

Figure 27.2 In the housing field rigid polyvinyl chloride siding, gutters, downspout, and soffit panels provide toughness, thermal insulation, and durability for years of service

applications, automobile seat covers, shower curtains, and cloth or paper coated sheet for upholstery materials and raincoats. Other applications include wire coatings, chemical storage tanks, gutters and siding for houses, packaging, and blow bottles which take advantage of its good clarity and toughness (Figures 27. 1 and 27. 2).

The majority of products made of PVC are processed by extrusion, blow molding, injection, rotational, and calendering. Fluidized bed coating, foaming, transfer and compression molding are also used for some of the compounds.

1. New words

vinyl['vainil] *n.* 乙烯基
formulation[fɔmju:'leiʃən] *n.* 配方
chloride['klɔ:raid] *n.* 氯化物
acetylene[ə'setli:n] *n.* 乙炔
extinguish[iks'tiŋgwiʃ] *v.* 熄灭
compounder[kɔm'paundə] *n.* 混炼机
gutter['gʌtə] *n.* 水槽
toughness['tʌfnis] *n.* 韧性
rotational[rəu'teiʃənl] *n.* 滚塑
transfer['trænsfə:] *n.* 压铸,传递
downspout['daunspaut] *n.* 水落管
soffit['sɔfit] *n.* 下端背面,拱腹
durability[ˌdjuərə'biliti] *n.* 经久,耐久力

2. Phrases and expressions

be derived from 来源于
polyvinyl chloride 聚氯乙烯
color possibility 着色性
self-extinguishing 自熄性
weather resistance 耐候性
abrasion resistance 耐磨性
flexible sheet 软片
rigid pipe 硬管
simulated leather 仿皮革
coated sheet 涂布片材
compression molding 压制模塑
transfer molding 传递模塑
thermal insulation 绝热

3. Notes to the text

① PVC is produced commercially from acetylene and hydrogen chloride and can be compounded to give almost any degree of flexibility to the final product by adding …
聚氯乙烯树脂的商业化生产是由乙炔气体与氯化氢单体聚合而成,在聚氯乙烯树脂中加入不同量的……即可制成各种柔韧性制品。

4. Exercises

(1) Prepare a chart showing the different ways in which PVC is used for products according to the properties it compounds.

(2) List four molded products that make use of the unique properties of PVC.

(3) Make a collection of small, natural color samples of polymers in granular or powder form.

(4) Make a list of the outstanding properties of plastics in general that makes them so valuable for commercial products.

(5) What is the main point of the second paragraph?

(6) Translate the following phrases into English:

性能范围	从硬制品到软制品	与其他塑料所不同的特点
硬管和软管	软片	侧基
仿革制品	汽车座罩	主要性能

Reading Material

Styrene Resins and Polystyrene

Styrene resins have been in use for many years and have proven to be polymers that the industry has kept improving by modifying and copolymerizing to up-grade their properties[1]. From the early brittle styrenes, industry now has a wide range of styrene based polymers with outstanding properties.

Polystyrene is a clear, odorless, and tasteless polymer whose structure is long chain, linear, and amorphous.

$$\left[CH_2 - \underset{\displaystyle C_6H_5}{CH} \right]_n$$

Produced from ethylene and benzene, the polymerized monomer forms a giant molecule with large side groups. It is isotactic. Scientists have worked with polystyrene structure to the extent that the polymer now produced possesses much better properties than the older resins[2]. In general terms, polystyrene can best be described as being crystal clear, rigid, and easy to process.

The wide melting range of polystyrene gives it versatility and ease of molding at various temperatures and pressures. However, there are two major disadvantages to polystyrene, its brittleness and poor chemical resistance. Brittleness is caused by the inflexibility of the molecule chains. This can be overcome by the physical addition of synthetic rubber, butadiene. This material is then known as high impact styrene and its strength is improved tremendously, making it a much more versatile plastic. The important properties of polystyrene are as follows:

(1) High degree of hardness;

(2) Brittle, except when modified;

(3) Excellent electrical properties;

(4) Holds static electricity and picks up dust;

(5) Good clarity and surface smoothness;

(6) Low moisture absorption;

(7) Ease of fabrication;

(8) Low cost;

(9) Clear and colorless, permits outstanding colorability in transparent, translucent and opaque shades;

(10) Poor outdoor weather resistance;

(11) Normal chemical resistance is good but softens on exposure to hydrocarbons like lacquer thinner.

Polystyrene products are found almost everywhere. Due to their low cost they are used for many disposable products such as picnic utensils, food containers and novelties. Typical molded products include refrigerator parts, appliance housings, furniture (Figure 27.3), automobile interior parts, plastic optical pieces, bottles and imbedded electrical parts. Extruded polystyrene sheet is a favorite for thermoforming packaging containers, lighted indoor signs and housewares. Injection mold model airplane and car kits make extensive use of polystyrene, especially because it is easily cemented. A more recent use of polystyrene is as an expanded foam. It is outstanding for packaging delicate parts (Figure 27.4), and as an insulation and floatation material. Picnic jugs and coolers take advantage of its outstanding insulation values.

Figure 27.3 Plastic has entered the furniture industry in large volume. The solid parts of this dining set were injection molded from polystyrene

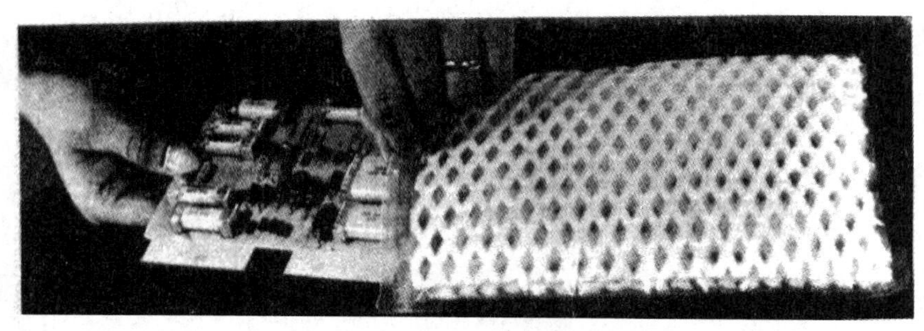

Figure 27.4 These delicate electronic components are safely packaged in expanded polystyrene foam for shipping

1. New words

modifying['mɔdifaiiŋ] *n.* 改性

butadiene[ˌbjuːtə'daIiːn] *n.* 丁二烯

picnic['piknik] *n.* 野餐

utensil[juː'tensl] *n.* 用具

novelty['nɔvəlti] *n.* 新产品

optical['ɔptikəl] *adj.* 光学的

model['mɔdl] *n.* 模塑

cement[si'ment] *n.* 黏合剂

expanded[iks'pændid] *adj.* 膨胀的 | delicate['delikit] *adj.* 精致的

2. Phrases and expressions

to the extent 在……程度上 | be caused by 起因于

3. Notes to the text

① Styrene resins have been in use for many years and have proven to be polymers that the industry has kept improving by modifying and copolymerizing to up-grade their properties. 虽然苯乙烯类树脂已经使用了许多年,但为了提高它们的性能,工业上采用改性和共聚的方法对其性能进行不断地改善。

② Scientists have worked with polystyrene structure to the extent that the polymer now produced possesses much better properties than the older resins. 科学家对聚苯乙烯结构进行研究,使现在生产出的聚合物比原来的树脂具有更好的性能。

Lesson 28 Styrene-Acrylonitrile(SAN) and Acrylonitrile-Butadiene-Styrene(ABS)

A copolymer of styrene, styrene acrylonitrile has improved properties of stiffness, chemical and scratch resistance and higher heat resistance. It is produced by the copolymerization of acrylonitrile and styrene to form the SAN polymer.

$$\left[CH_2 - \underset{C_6H_5}{CH} - \underset{C\equiv N}{CH} \right]_n$$

Due to the copolymerization, the water-white color of polystyrene is changed to a slight yellow cast in SAN. Stress crack resistance is improved. SAN finds applications in decorative panels, food packages, tumblers (Figure 28.1), lenses, batteries, and telephone parts and piano keys. However ease of fabrication and cost lose out to gain the better properties. Styrene acrylonitrile may be molded and fabricated by processes used with polystyrene.

Figure 28.1 Drinking cups made of styrene acrylonitrile provide good insulation for hot or cold beverages and adequate impact strength

Three monomers, acrylonitrile, butadiene, and styrene are used to produce the ABS TERPOLYMER, (TER) meaning three. It is a further way in which the properties of polystyrene can be further enhanced over SAN[①]. Unlike the physical addition of butadiene to styrene to produce a higher impact material, ABS is a chemically polymerized plastic.

$$\left[\left(CH_2 - \underset{CN}{CH} \right)_x \left(CH_2 - CH = CH - CH_2 \right)_y \left(CH_2 - \underset{C_6H_5}{CH} \right)_z \right]_n$$

ABS can be described as a rugged, tough plastic with moderately good chemical resistance and a high heat distortion point. It is one of the few thermoplastics which combines both hardness and toughness. As with most others, if it is hard, it is brittle; if it is tough, it is flexible. In its natural state ABS is a light tan colored opaque plastic with the following major properties:

(1) Will withstand temperatures up to 212 ℉[②].

(2) Low coefficient of friction.

(3) Good wear and scratch resistance.

(4) Resistant to most common chemicals and some hydrocarbons.

(5) Good electrical properties, but flammable.

(6) High hardness and rigidity.

(7) Remains tough at −40 ℉.

(8) Good colorability except for transparency.

The ABS polymers can be processed through most molding processes including calendering and rotational molding. They are available mainly as powders and granules ready for processing. Typical products made of ABS are vacuum formed refrigerator door liners, luggage cases, and boat hulls. Extruded pipe and pipe fittings have found many applications. Other typical products are football helmets, a variety of automobile trim and hardware, telephone and power tool housings, tool handles, gears, and radio and television cases (Figures 28.2 and 28.3).

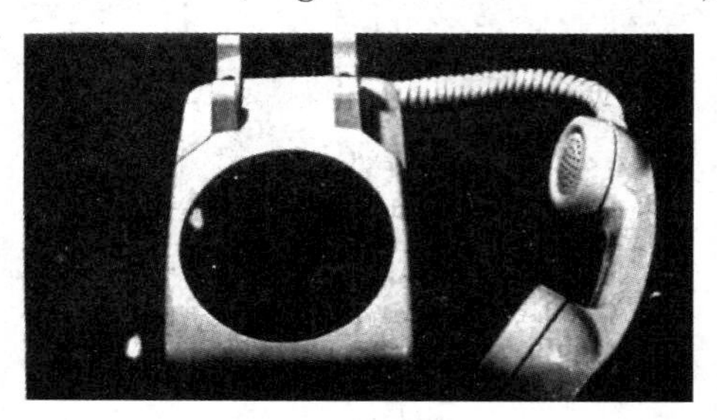

Figure 28.2 The receiver and base of this telephone were injection molded from ABS resin

Figure 28.3 ABS resin was used to mold this complete automobile body. The properties of ABS make it ideal for automotive uses

1. New words

terpolymer[tə:'pɔlimə] *n.* 三元共聚物
enhance[in'ha:ns] *v.* 提高,增加
moderately['nɔdəritli] *adv.* 适中地
tan[tæn] *n.* 棕褐色
opaque[əu'peik] *adj.* 不透明的
wear[wεə] *v.* & *n.* 磨损
transparency[træns'pεərənsi] *n.* 透明性
helmet['helmit] *n.* 头盔
hardware['ha:dwεə] *n.* 金属构件

2. Phrases and expressions

chemically polymerized 化学聚合
heat distortion point 热扭变点
wear resistance 耐磨性
ready for 预备好

3. Notes to the text

① It is a further way in which the properties of polystyrene can be further enhanced over SAN. 这是比 SAN 更进一步提高聚苯乙烯性能的方法。

② Will withstand temperatures up to 212℉. 耐温高达 212℉。

4. Exercises

(1) SAN copolymer is produced by the copolymerization of ________ and ________.

(2) ABS is one of the few plastics which combines both the properties of ________ and ________.

(3) What are the major properties of ABS?

(4) Translate the first two paragraphs into Chinese.

Reading Material

Polyurethane

Polyurethane resins are produced through polymerization of isocyanate and hydroxyl groups and are given the name "isocyanates" as a family of polymers.

$$\left[\overset{\overset{\displaystyle O}{\|}}{C} - NH - CH_2 - CH_2 - NH - \overset{\overset{\displaystyle O}{\|}}{C} - O - CH_2 - CH_2 - O \right]_n$$

Polyurethane is somewhat like the vinyls in that it can be prepared as a rigid molding material or a flexible, rubbery material. The major portion of polyurethane goes into the production of foamed plastics but the use of the rigid polymer is on the increase.

Rigid polyurethane molding granules are an amber translucent color. They are called elastomers as they will stretch to more than twice their size and return to their original shape[①]. These elastomers are extremely abrasion resistant, very tough, and resist tear and shock. They are chemically resistant to almost all common chemicals including oils, solvents, and acids. Rigid polyurethane elastomers remain flexible down to about $-40\ ^\circ F$. They have good electrical properties and high load bearing capacity (Figure 28.4).

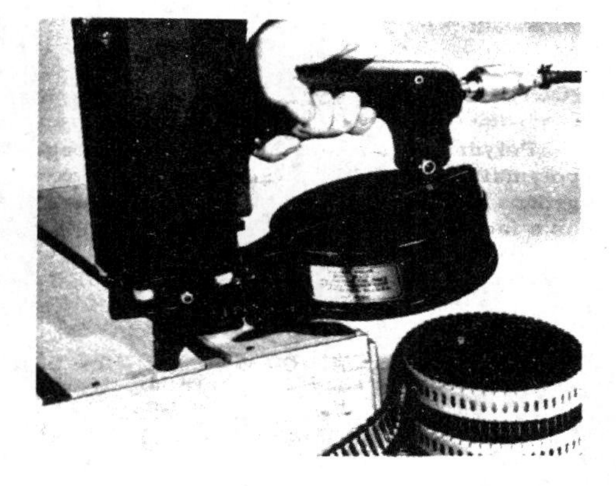

Figure28.4 The magazine of this automatic mailer made of polyurethane withstands the high impact of driving 18-mails per minute

Foamed polyurethane has similar properties to the rigid material. It can be produced either as rigid or flexible foams in an open cellular structure. Flexible foams are good, sound and energy absorbers, providing good cushioning properties.

Due to the rubber-like properties of the rigid polyurethane polymers and their extreme toughness, they are used as solid tires on heavy equipment, printing and materials handling rolls, gaskets, bumpers and shock impact devices (Figure 28.5), and synthetic

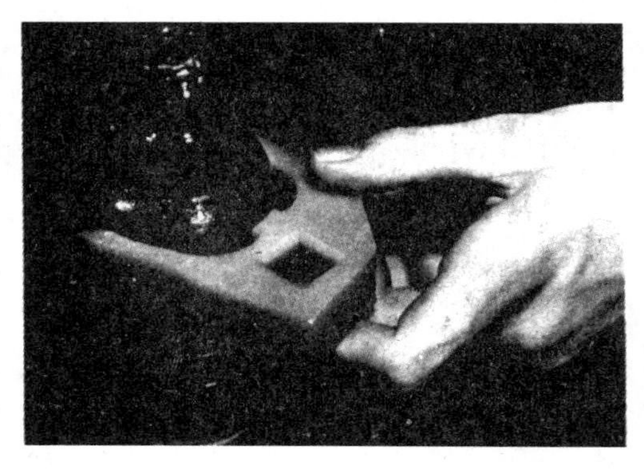

Figure 28.5 Bottles are picked up along the conveyor line by this flexible polyurethane finger grip and moved to the filling area

leather. The rigid materials are injection molded, extruded, coated and cast.

Polyurethane foams are available as two component spray units for coatings, as liquids which are foamed in molds or foamed-in-place, and as slab stock for fabricating②. These foamed materials are used in refrigerator insulation, sponges, crash pads, automobile and furniture cushioning, and cavity filling in boats, airplane wings, pontoons, and life jackets.

1. New words

isocyanate[ˌaisəu'saiəneit] *n.* 异氰酸酯
hydroxyl[hai'drɔksil] *n.* 羟(基)
tear[tɛə] *v.* & *n.* 撕裂
shock[ʃɔk] *n.* 冲击,振动
cellular['seljulə] *adj.* 多孔的 泡沫的
cushion['kuʃən] *v.* 缓冲
pontoon[pɔn'tuːn] *n.* 浮桥(船)

2. Phrases and expressions

hydroxyl group 羟基
resist tear 耐撕裂
down to 降到
load bearing capacity 承载能力
open cellular structure 开孔结构，开放的网格结构
foamed-in-place 现场发泡
slab stock 泡沫塑料，块料
crash pad 防震垫
life jacket 救生衣

3. Notes to the text

① They are called elastomers as they will stretch to more than twice their size and return to their original shape. 可将硬聚氨酯称为弹性体,因为将它们伸展两倍尺寸后仍能恢复原状。

② Polyurethane foams are available as two component spray units for coatings, as liquids which are foamed in molds or foamed-in-place, and as slab stock for fabricating. 聚氨酯泡沫塑料制品的类型主要有:两组分喷涂的涂层制品,液体模塑或现场发泡制品及二次加工的块状制品。

Lesson 29 Polycarbonate

There are few thermoplastics that have the outstanding engineering properties of polycarbonate. In recent years it has taken its place among the most valuable resins of the industry. Polycarbonate is a member of the polyester family in that it contains carbon and oxygen atoms as the backbone of the molecular chain.

$$\left[O - C_6H_4 - C(CH_3)_2 - C_6H_4 - O - \underset{\underset{O}{\|}}{C} \right]_n$$

The plastics industry is using polycarbonate in direct competition with nylon and acetal as well as metals like copper, zinc, and brass.

One of the toughest of all plastics, polycarbonate meets many of the extreme properties of plastics with little loss in general characteristics[1]. It is available in granular and sheet form ready for molding or fabricating. This polymer is a water-clear material, easily colored and is adaptable to a variety of molding processes including injection, extrusion, and blow molding. Five outstanding properties of polycarbonate which, in combination, separate it from other thermoplastics materials are:

(1) Good electrical properties which meet the needs of electrical and electronic industries.

(2) Outstanding impact strength which makes it useful for long service life under extreme, heavy-duty conditions[2] (Figure 29.1).

(3) The transparency of polycarbonate approximates that of the light transmission of the acrylics and glass.

(4) Superior dimensional stability qualifies it for use in precision-engineered components where close tolerances are required[3].

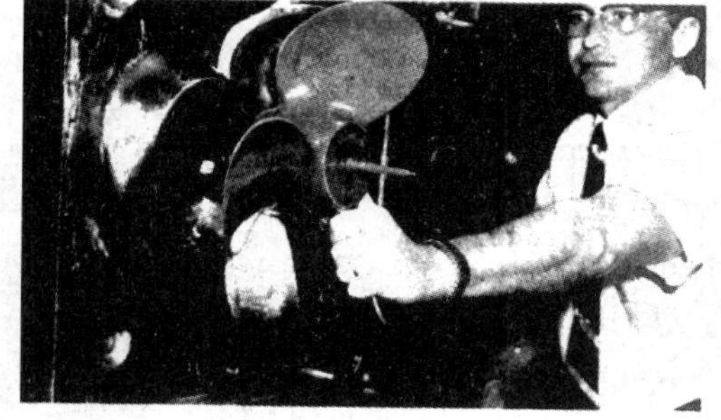

Figure 29.1 Injection molded polycarbonate boat propeller. The toughness and flexibility of polycarbonate are causing designers to consider it for replacement of many metal parts

(5) The self-extinguishing properties of polycarbonate make it useful in applications involving high temperature use where safety hazards may exist.

Polycarbonate also contains the average properties of many other plastics. These include good machinability, high temperature stability, good weatherability and good chemical resistance. It is attacked by some hydrocarbons and can be dissolved in ethylene dichloride.

The outstanding properties of polycarbonate make it suitable for applications where other thermoplastics are inadequate. Polycarbonate street lighting globes are easily blow molded and are not brittle like glass(Figure 29.2). This advantage shows up in products like protective face masks, covers for electrical panels, electrical insulators and window panes for buildings. Other applications of polycarbonate include football and safety helmets, sunglass lenses, blow molded bottles, shoe heels, electric can openers, coffee pots, and housings for shavers, power tools and air conditioners.

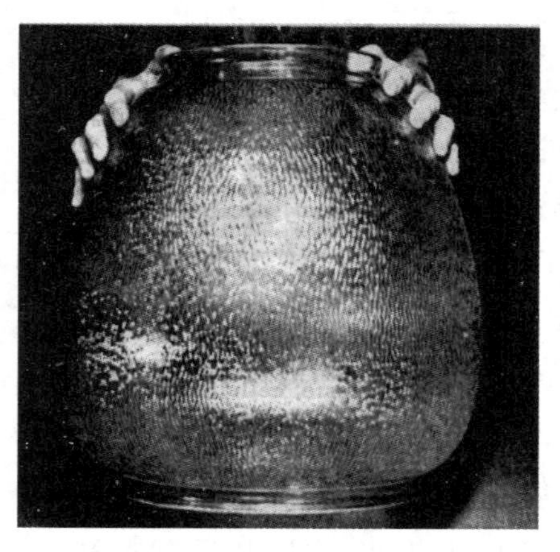

Figure 29.2 Blow molded street lighting globes made of polycarbonate provide good transparency and considerable resistance to breakage

1. New words

polycarbonate [ˌpɔliˈka：bənit] *n.* 聚碳酸酯
dimensional[diˈmenʃənl] *adj.* 尺寸的，有尺寸的
dichloride[daiˈklɔ：raid] *n.* 二氯化物
globe[gləub] *n.* 球体，球形物
tolerance[ˈtɔlərəns] *n.* 公差
machinability[məˈʃi：nəbiliti] *n.* 机械加工性

2. Phrases and expressions

in competition with 与……竞争
in combination 结合
separate…from 把……从……中分开
precision-engineered components 工程零件
ethylene dichloride 二氯化乙烯
electrical panel 配电板罩

3. Notes to the text

① One of the toughest of all plastics, polycarbonate meets many of the extreme properties of plastics with little loss in general characteristics. 聚碳酸酯是所有塑料材料中韧性最好的材料之一，一般说来它具有塑料的许多优良性能而缺点很少。

② Outstanding impact strength which makes it useful for long service life under extreme, heavy-duty conditions. 聚碳酸酯的冲击强度十分优良,能长期在极端或受载荷的条件下使用。

③ Superior dimensional stability qualifies it for use in precision-engineered components where close tolerances are required. 聚碳酸酯具有较高的尺寸稳定性,可以作为精密配合的工程零部件使用。

4. Exercises

(1) Polycarbonate is a member of the ________ family.

(2) List five outstanding properties of polycarbonate which separate it from other thermoplastic materials.

(3) What is meant by self-extinguishing?

(4) What does the word "dimensional stability" in the third paragraph mean?

(5) List three major ways in which polycarbonate resin is processed into products.

Reading Material

Nylon

Nylon is the common name for polyamide resins. This is a group of complicated, long chain molecule polymer derived from amino and other acids.

$$\mathrm{+\!\!\!\![NH(CH_2)_m—NHCO—(CH_2)_{n-2}—CO]\!\!\!\!+_x}$$

Nylon was first introduced by the Du Pont Company in 1938 and quickly became well known when it was used as fibers to weave hosiery as a replacement for silk. Because of its high impact strength, versatility, and ease of processing, nylon has become one of the leading polymers of the plastics industry. A number of modifications in production practices can be made which result in different degrees of flexibility and stiffness①. This affords considerable variations, and enables specific tailor-made resins to be produced for many applications.

Nylon is translucent, off-white in color, and has a high surface gloss. It can be easily colored to a broad range. Nylon is a sensitive material to process since it has a narrow melting range and must be dried, as it absorbs moisture quickly. The outstanding properties of nylon can be characterized by high heat and chemical resistance and by outstanding toughness and flexural strength (Figure 29.3). Further properties include:

(1) High abrasion resistance.

(2) Low coefficient of friction.

(3) Good resistance to chemicals; hydrocarbons and oils, but attacked by strong acids.

Figure 29.3 Automotive fuse block made of nylon to reduce high rate of breakage in mounting basses and legs during auto assembly

(4) High moisture absorption causes dimensional change.

(5) Excellent water resistance.

(6) Fair electrical properties.

Nylon is available in granules as a molding material, as a powder for coatings, and in standard sheet, rod, tubing, and fibers②.

Nylon continues to find new applications in products such as electric tool housings and shaver cases where designers can take advantage of its strength in thin wall sections. Automotive interior light covers are making use of its natural translucency. Nylon is injection molded, extruded, blow molded, and used in powder molding processes for bearings, gears (Figure 29.4), hinges, drawer slides and rollers, combs, ship propellers, fishing lines, and textiles. More small industrial parts are made of nylon than most other plastics.

Figure 29.4 Long wearing, quiet gears made of nylon require no lubrication. Gears in this gear train range in diameter from 2 3/4in. to 8ft

1. New words

amino['æminəu] *adj.* 氨基的

sensitive['sensitiv] *adj.* 敏感的

abrasion[ə'breiʒən] *n.* 磨损,磨蚀

bearing['bɛəriŋ] *n.* 轴承

gear[giə] *n.* 齿轮

hinge[hindʒ] *n.* 铰链

propeller[prə'pelə] *n.* 螺旋桨

textile['tekstail] *n.* 纺织材料

lubrication[ˌlu:bri'keiʃən] *n.* 润滑油

2. Phrases and expressions

derived from 从……中产生

Du. Pont Company 杜邦公司

fuse block 保险丝装置,熔丝盒
tailor-make 特制的,适合的
off-white 纯白色的, 米色的
surface gloss 表面光泽度
flexural strength 挠曲强度
strong acid 强酸
water resistance 耐水性

3. Notes to the text

① A number of modifications in production practice can be made which result in different degrees of flexibility and stiffness. 在生产实践中,对尼龙进行改性,可得到具有不同程度柔性和刚性的尼龙材料。

② Nylon is available in granules as a molding material, as a powder for coatings, and in standard sheet, rod, tubing, and fibers. 粒状尼龙可以注塑成型,粉状尼龙可以涂层,也可以成型为标准的片材、棒材、管材和纤维材料。

Lesson 30　Epoxy Resins

The epoxy resins have steadily grown in their use in the plastics and related industries due to the variety of forms in which they may be processed. As thermosetting materials, they are cured or cross-linked by the addition of a hardener to the original liquid resin[1]. The repeating molecular structure is attached to terminal molecular groups as curing takes place[2].

$$\underset{\text{O}}{\overset{}{CH_2\text{—}CH}}\text{—}CH_2\text{—}\left[O\text{—}C_6H_4\text{—}\underset{CH_3}{\overset{CH_3}{\underset{|}{\overset{|}{C}}}}\text{—}C_6H_4\text{—}O\text{—}CH_2\text{—}\underset{OH}{\underset{|}{CH}}\text{—}CH_2\right]_n\text{—}O\text{—}C_6H_4\text{—}\underset{CH_3}{\overset{CH_3}{\underset{|}{\overset{|}{C}}}}\text{—}C_6H_4\text{—}O\text{—}CH_2\text{—}\underset{\text{O}}{CH\text{—}CH_2}$$

Epoxy resins are formulated in such a manner that they are adaptable to many processing techniques. In each of these, the outstanding characteristics of the epoxies provide excellent chemical resistance, electrical properties, and toughness. Most epoxy products can be used continuously at temperatures up to 300 ℉ while special formulations with fillers and additives resin their properties at continuously elevated temperatures up to 500 ℉. They have excellent mechanical and thermal shock resistance except for some rigid formulations which are quite brittle. Weather resistance and low temperature properties are good, even to temperature at −70 ℉.

Figure 30.1　Glass fibers impregnated with epoxy resin formed the outer casing for the Hercules third stage rocket motor and payload

Extensive use is made of epoxy resins in coating systems. This includes coating for corrosion and abrasion resistance in containers, pipe and tank liners, floor and wall finishes, steel and masonry surfaces. Epoxy adhesives are very strong and are especially adaptable to metals, glass, ceramics and dissimilar materials. Molding compounds of epoxy are available with catalysts incorporated ready for compression and transfer molding into such product as pipe fittings, electrical components, and bobbins for coil winding. The epoxies are used in a manner similar to the polyesters as glass fiber reinforced lay-ups, and as laminated sheet material(Figure 30.1). Cast epoxy is used for short run molds, tools, and jigs, being filled up to 50 percent with powdered aluminum or other inert binders. Potting and encapsulation(coating with plastic for protection) is used for electronic parts, bushings, and insulators. Other applications include printed circuit boards, boat bodies, aircraft skins, body solders and sealers.

1. New words

epoxy[e'pɔksi] *adj.* 环氧的
addition[ə'diʃən] *n.* 加成作用
hardener['ha:dnə] *n.* 固化剂
terminal['tə:minl] *n.* 端基
corrosion[kə'rəuʒən] *n.* 腐蚀
masonry['meisənri] *n.* 建筑
binder['baində] *n.* 黏合剂
encapsulation[inˌkæpsju'leiʃən] *n.* 封铸,封装

2. Phrases and expressions

epoxy resins 环氧树脂
thermal shock resistance 耐热骤变性

3. Notes to the text

① As thermosetting materials, they are cured or cross-linked by the addition of a hardener to the original liquid resin. 作为热固性材料,它们是通过将固化剂加到液体树脂中而发生固化或交联作用的。

② The repeating molecular structure is attached to terminal molecular groups as curing takes place. 当固化反应发生时,重复的分子链节即被连接到端基分子基团上了。

4. Exercises

(1) Weather resistance and low temperature properties of epoxy resins are good, even to temperature at ________.

(2) List three special materials that epoxy adhesives can bind them together.

(3) Make a bulletin board display of plastic polymer ads.

(4) Secure a number of different throw-away plastic containers. Make a list of the plastic property requirements necessary to make each container function properly.

(5) Translate the following sentences into Chinese and pay attention to the translation of the words or structures used for emphasis.

a. This reaction did take place.

b. It is to accomplish this operation with the minimum expenditure of energy that is the principal concern of extrusion process designer.

c. It was not until 1920 that the macromolecular hypothesis could be accepted.

d. Hence comes the name elastomer.

Reading Material

Tetrafluoroethylene Resins

Tetrafluoroethylene is the major member of the family of fluorocarbon polymers. It is a close relative of polyethylene, as a paraffin hydrocarbon, in which all of the hydrogen atoms are replaced with fluorine atoms①. It is a long chain, linear polymer with extremely high crystallinity, 93 ~ 97 percent. The unique properties of tetrafluoroethylene and its availability in a wide range of forms make it one of the most valuable plastics on the market.

$$\left[CF_2—CF_2 \right]_n$$

Figure 30.2 Excellent electrical properties make tetrafluoroethylene well suited for use as a wire coating. Wires with thin coatings require minimum of space

Tetrafluoroethylene has many distinguishing properties. It is the most inert (chemically inactive) of all plastics, resisting attack from most every chemical compound known even at high temperatures. The coefficient of friction is the lowest of any known solid material. When it is exposed to temperature changes, it is still flexible at -450 °F and stable up to 500 °F. The molecular weight of the polymer is unusually high, perhaps reaching several million. The natural color of tetrafluoroethylene is an opaque white which is readily colored. The polymer is extremely tough, possesses excellent electrical properties (Figure 30.2), and has a waxy feeling to the touch. The low coefficient of friction also accounts for the no-stick properties displayed in many products.

Tetrafluoroethylene is a difficult material to process due to its high melting point and poor flow characteristics. Processing is accomplished by dip coating of TFE dispersions, extruding lubricated powders, and sintering similar to powdered metals②. Sintering powdered TFE consists of compressing the material at room temperature and high pressure to a solid form. The formed material is then heated to about 700 °F at which time fusion takes place forming a solid mass.

Figure 30.3 This tube of tetrafluoroethylene is to be heat shrunk on the distributor roll of a printing press. With this type roll nonsticking and clean-up qualities are improved

Many industrial and consumer products

make use of the extreme properties of tetrafluoroethylene in applications such as no-stick cookware, electrical insulation, chemically resistant gaskets, piston rings, bearings and tubing(Figure 30. 3). Test tubes and other containers of TFE find use in chemical laboratory ware. The most commonly seen use of TFE is as a non-stick coating on rolling pins, irons, frying pans and other household appliances.

1. New words

tetrafluoroethylene[ˌtetrəˌfluərə'eθiliːn] *n.* 四氟乙烯
paraffin['pærəfin] *n.* 石蜡,烷烃
fluorine['fluəriːn] *n.* 氟
availability[əveil'əbiliti] *n.* 适用性,使用价值
distinguish[dis'tiŋgwiʃ] *v.* 特性,区分
inert[i'nəːt] *adj.* 惰性的
inactive[in'æktiv] *adj.* 不活泼的
stable['steibl] *adj.* 稳定的
dispersion[dis'pəːʃən] *n.* 分散
lubricate['ljuːbrikeit] *v.* 上油,使润滑
sinter['sintə] *n.* & *v.* 烧结
fusion['fjuːʒən] *n.* 熔结,熔融
piston['pistən] *n.* 活塞
distributor[dis'tribjutə] *n.* 发行人,分散

2. Phrases and expressions

fluorocarbon polymer　氟碳高聚物
chemically inactive　化学惰性
dip coating　蘸涂
sintering powdered　粉料烧结的
no-stick cookware　不黏炊具
distributor roll　分散辊,(油墨)匀布辊

3. Notes to the text

① It is a close relative of polyethylene, as a paraffin hydrocarbon, in which all of the hydrogen atoms are replaced with fluorine atoms.　像聚乙烯一样四氟乙烯也是一种饱和链烃,只不过其中所有的氢原子已经被氟原子所置换。

② Processing is accomplished by dip coating of TFE dispersions, extruding lubricated powders, and sintering similar to powdered metals.　四氟乙烯的加工可采用TFE分散蘸涂成型、粉末挤压成型及类似于粉末金属的冷压烧结成型。

Lesson 31 Thermosetting Materials and Phenol Formaldehyde Resins

Thermosetting Materials:

Thermosetting materials differ from thermoplastics in many ways. Basically the differences are due to the chemical condensation polymerization process through which the thermosetting materials are produced that results in their special characteristics[1]. The definition of thermosetting polymers illustrates that they are a cross-linked network of long chain molecules causing them to be rigid, strong, and infusible. This means that once they have become set or molded into a given shape they cannot be reheated and reshaped. Thermosets also have two other characteristics that are different from thermoplastics. They are limited in the number of processes by which they can be molded and scrap resulting from molding or reject parts cannot be reused. There are, however, many advantages to thermosetting materials which will be brought out in the descriptions of the individual polymers. Most all of the thermosetting materials have fillers added to improve their properties and to extend the volume of the resin. Typical fillers are wood flour and glass fibers.

Phenol Formaldehyde Resins:

The phenol formaldehyde resins, more commonly known as phenolics, are produced from the reaction of phenol (carbolic acid) with formaldehyde in the presence of a catalyst[2]. The resulting polymer can be obtained as a liquid for casting, bonding, coating, and impregnating, or as molding compounds as powders or flakes. Phenolic molding compounds are available only in dark colors while casting and impregnating resins can be obtained in relatively clear solutions.

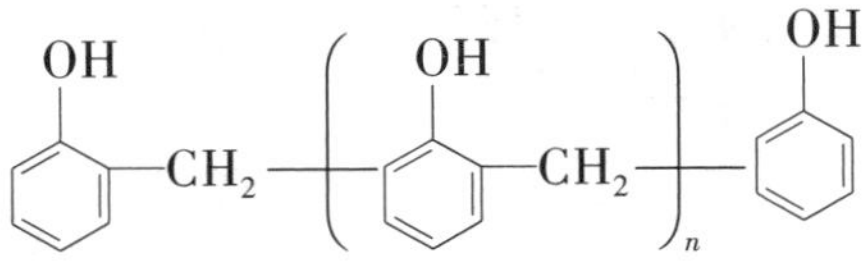

OH OH OH

CH_2 CH_2 n

($n = 4 \sim 12$)

Phenolics can best be described as being hard, rigid, heat resistant materials that are quite brittle unless they are filled. They are seldom used in molded products without fillers which improve their toughness. Characteristics of phenolics that make them so valuable to the plastics industry are their relatively low cost, excellent insulating properties, heat resistance to 500 °F, and chemical inertness to most common solvents and weak acids. The dimensional stability and low moisture absorption give them an advantage in the design of precision devices.

Phenolics are processed by compression and transfer molding. They are also used as liquids in laminating of veneers, fabrics, and paper, and in coating and adhesive application. Some typical product uses of phenolics are distributor caps and coil tops, impregnated brake linings, telephones, tool housing (Figure 31.1), and home appliance handles and parts. Considerable volume is used as an adhesive in bonding plywood. Impregnated wood fibers are molded into salad bowls and croquet balls (Figure 31.2), while reinforced laminated sheet is used for roof panels and automobile body parts. Phenolic bonding agents are used in the foundry industry for shell molding and cores.

Figure 31.1 Tool housings, such as this power drill, make use of the rigidity, strength, and insulating properties of molded phenolic

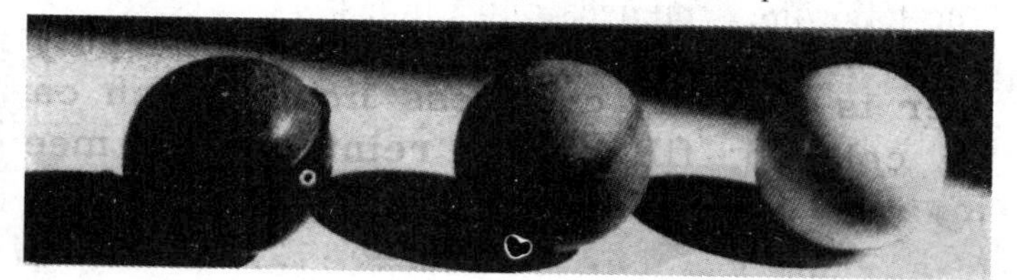

Figure 31.2 Tough, durable croquet balls made from phenolic resin mixed with wood particles and compressed under high heat and pressure

1. New words

reject['ri:dʒekt] *n.* 废料
carbolic['kɑ:bəlik] *adj.* 碳的
impregnate['impregneit] *n.* 浸渍树脂; *v.* 浸渍
flake[fleik] *n.* 片状
precision[pri'siʒən] *n.* 精密度
laminating['læmineitiŋ] *n.* 层压,层合
veneer[və'niə] *n.* 板坯
distributor[dis'tribjutə] *n.* 配电盘
foundry['faundri] *n.* 铸造厂
core[kɔ:] *n.* 模芯,阳模
croquet['krəukei] *n.* 槌球戏,循环球戏

2. Phrases and expressions

condensation polymerization process 缩聚方法,缩聚过程
cross-linked network 交联网状
brought out 引出,生产,阐述

laminated sheet 层压板 | power drill 机械钻

3. Notes to the text

① Basically the differences are due to the chemical condensation polymerization process through which the thermosetting materials are produced that results in their special characteristics. 本质的区别是通过化学缩聚法产生的热固性材料具有特殊的性能。

② The phenol formaldehyde resins, more commonly known as phenolics, are produced from the reaction of phenol(carbolic acid) with formaldehyde in the presence of a catalyst. 苯酚甲醛树脂通称为酚醛塑料,是由苯酚(碳酸)与甲醛在催化剂的作用下进行反应得到的。

4. Exercises

(1) Most thermosetting plastics contain fillers to ________ and to ________ .

(2) Four ways in which the properties of phenolics can best be described are ________ .

(3) Characteristics of phenolics that make them so valuable to the plastic industry are their ________ .

(4) Write a brief report explaining why thermosetting and thermoplastic polymers are seldom used in the same processing equipment.

(5) Translate the following sentences into Chinese:

a. Thermosets have not yet been polymerized before processing, and the chemical reaction takes place during the process, usually through heat, a catalyst, or pressure.

b. Injection molding machines are manufactured in many sizes, and these are rated according to size by the amount of material which can be injected in one cycle, which ranges from a fraction of an ounce in the small laboratory models to many pounds in large production equipment.

Reading Material

Polyester

Polyester is best known as the resin combined with glass mat or chopped fibers and called fiber glass. The polymer is produced from the polymerization of certain alcohols and acids. It can take the form of completely polymerized polyester film or as a resin to which a catalyst is added to complete the curing during molding①. The polymer is a clear, colorless liquid which can be colored, filled, and reinforced to meet many specifications.

$$\left[O-CH_2-CH_2-O-\underset{\underset{O}{\|}}{C}-C_6H_4-\underset{\underset{O}{\|}}{C}-O-CH_2-CH_2 \right]_n$$

The properties of polyesters are varied as to the form in which they are processed②. In general they have very good weathering and chemical resistance properties, withstanding most solvents, acids, and salts. Polyesters are quite strong and tough, and depending upon the materials used for reinforcing, can be made rigid or fairly flexible. The electrical insulation properties are high along with good heat resistance to 500 ℉ and colorability. Polyesters are available as liquid resins along with catalysts which are mixed at the time of use. Premixed molding compounds with fillers are available for immediate use. These compounds will react if not used quickly or kept cold.

Polyester laminating and molding with reinforced fibers is one of the largest commercial processes used for product manufacture. Matched die, bag, or hand lay-up molding is used to produce such items as boat hulls, automobile and aircraft body components (Figure 31. 3), translucent sheets for roofing (Figure 31. 4), wash tubs, and luggage. Premix molding produces such products as hammer handless and automotive ductwork by the compression process. Cast polyesters are used for embedding scientific specimens, decorative jewelry, and cast-in electrical parts. Polyesters are also used as coating on masonry, wood, and metal and as patching compounds.

Figure 31. 3 Polyester reinforced resins of high strength, fatigue resistance, fire resistance and lightweight make an ideal outer body for this gyroplane

Figure 31. 4 Translucent polyester reinforced roofing panels provide interior lighting combined with lightweight and good weather resistance

1. New words

cure[kjuə] *n.* 固化,硫化
specification[ˌspesifiˈkeiʃən] *n.* 技术要求
embedding[imˈbediŋ] *n.* 镶铸
gyroplane[ˈdʒaiərəplein] *n.* 旋翼机
translucent[trænzˈljuːsənt] *adj.* 半透明的,透明的

2. Phrases and expressions

glass mat 玻璃毡片
chopped fiber 碎纤维
premixed molding 预混模制
matched die 对模

compression process　压塑加工
fatigue resistance　耐疲劳
fire resistance　耐火

3. Notes to the text

① It can take the form of completely polymerized polyester film or as a resin to which a catalyst is added to complete the curing during molding.　聚酯树脂可以是完全聚合而成的聚酯膜,也可以是加入了催化剂以便在模塑时完成固化成型的树脂。

② The properties of polyesters are varied as to the form in which they are processed.　聚酯性能随加工形式的不同而不同。

Lesson 32 The Historical Development of Rubber

Of all the materials provided by nature for man to use as a material of construction, natural rubber is unique. The characteristic of high reversible extensibility fascinates the enquiring mind. As the years have passed since its introduction to the Eastern Hemisphere by Columbus and his fellow explorers the roll of famous men who have studied the material has steadily lengthened. These include Priestley, who coined the word 'rubber'; Faraday, who determined its empirical formula; Weizmann, who in later years became the first president of Israel, and the Nobel Chemistry Prizewinners Staudinger, Flory, Ziegler and Natta.

At the same time there is more than a touch of alchemy imposed on the magic of rubber. The process of mastication was discovered accidentally by Hancock in the 1820s. Only in the late 1940s was it demonstrated that this was due to the rupture of chemical bonds by mechanical means, a discovery which established the science of mechano-chemistry. The discovery of vulcanization, the process that renders rubber less temperature sensitive and also insoluble as well as regenerating high elasticity in masticated rubber, was made totally without understanding of the processes involved. The heating of pieces of rubber in pots of boiling sulphur by Goodyear and others in the early years of the 19th century seems more appropriate to the witches' brews of Shakespeare's Macbeth①. Nevertheless some 150 years later sulphur is still the monarch of vulcanizing agents.

There is a further wonder in that the source of natural rubber is a milky sap to be found in many hundreds of species of plants②. This milky sap, or latex, when coagulated and dried yields a material which is not only rubbery but in a way crystalline. Today there are many man-made rubbers but the natural product still plays a substantial role on the world's industrial stage, a story very different from that of many other natural materials challenged by synthetics.

On first consideration it may be thought that rubberiness is an invariant property, that rubber is rubber and that is all there is to it. On second thoughts one realizes that the number of products that can be made from natural rubber is very large and that the properties of such products may be very different. One has only to think of a type, a toy balloon, an ebonite battery box, an eraser, and a hot-water bottle to appreciate that considerable modification to the rubber may be made by judicious compounding. Indeed it may be argued that greater variation is possible by the selection of additives to the base rubber than by changing the base rubber.

Both raw and masticated rubber were however unsatisfactory for all but a few end uses. The key discovery by the American, Charles Goodyear, in 1839, that heating a mixture of rubber, sulphur and white lead gave a material of far superior properties, is without doubt the most important milestone in the history of rubber. This process, commercially exploited by Hancock, provided the public with an elastic material less liable to become sticky in hot weather and stiff in cold and also insoluble in common solvents. Another of Hancock's friends, Brockeden, termed the process vulcanization a term used to this day by the rubber technologist and chemist. Today we recognize that this process involves the cross-linking of polymer chains to form network structures which severely limit the ability of the chains to slide past each other③. Whilst we now know much more about the requirements of an ideal cross-linking agent and the mechanisms of these reactions, sulphur remains the dominant cross-linking agent and of the world's rubber industries.

Whilst Faraday had shown in 1826 that rubber was a hydrocarbon of empirical formula C_5H_8 few further important developments occurred until the closing years of the 19th century when the structure of natural rubber began to be progressively revealed, a process which only became substantially complete about 1930.

1. New words

Columbus[kə'lʌmbəs] *n.* 哥伦布
alchemy['ælkəmi] *n.* 炼丹术
mastication[ˌmæsti'keiʃən] *n.* 塑炼,捏炼
vulcanization[ˌvʌlkənai'zeiʃən] *n.* 硫化
pot[pɔt] *n.* 罐
witch[witʃ] *n.* 女巫
brew[bru:] *n.* 酿造,煎药
Macbeth[mək'bəθ] *n.* 麦克佩斯
monarch[mɔk'nək] *n.* 主要,统治,君主
milky['milki] *adj.* 乳状的
sap[sæp] *n.* 树汁
invariant[in'vɛəriənt] *adj.* 不变的
judicious[dʒu:'diʃəs] *adj.* 明智的
milestone['mailstəun] *n.* 里程碑

2. Phrases and expressions

empirical formula 经验化学式
mechano-chemistry 力化学
vulcanizing agents 硫化剂
white lead 铅白
liable to 易于……的,有……倾向的
cross-linking agent 交联剂

3. Notes to the text

① The heating of pieces of rubber in pots of boiling sulphur by Goodyear and others in the early years of the 19th century seems more appropriate to the witches' brews of Shakespeare's Macbeth. 19世纪初,Goodyear和其他人在沸腾硫黄罐中将橡胶片加热的情况,与莎士比亚《麦克佩斯》一剧中的女巫煎药的情形颇为相似。

② There is a further wonder in that the source of natural rubber is a milky sap to be found in many hundreds of species of plants. 更令人惊异的是,天然橡胶的源泉是数百种

植物的乳状树汁。

③ Today we recognize that this process involves the cross-linking of polymer chains to form network structures which severely limit the ability of the chains to slide past each other. 现在我们认识到,这一过程是聚合物链交联形成的网络结构,它严格限制了链的互相滑移能力。

4. Exercises

(1) Why is natural rubber unique among all materials of construction?
(2) Who introduced natural rubber to the Eastern Hemisphere?
(3) When was the process of mastication discovered?
(4) According to the passage what kinds of products can be made by rubber?
(5) Translate the last paragraph into Chinese.
(6) Make a list of the newest rubber products available in your home.

Reading Material

Synthetic Rubber

Long before the structure of rubber had been established attempts to prepare a synthetic rubber had been made. In 1860 Greville Williams distilled rubber and then fractionated the distillate. From the lowest fraction Williams obtained a liquid boiling at 37 ~ 38℃ which he called isoprene. Williams further described an experiment involving the exposure to air of isoprene which eventually led to the production of a 'white, spongy elastic mass.' Conceivably this was the first man-made rubber although the 'starting material' isoprene had itself been obtained from nature rubber.

In 1879 F. G. Bouchard produced a material from isoprene which clearly had many properties akin to the natural product but as with Williams' experiments the isoprene had been obtained from natural rubber. By 1884 W. A. Tilden had prepared isoprene from turpentine and then converted the former to a rubber-like material by methods similar to those used by Bouchard. At last a rubber had been made by sources independent of natural rubber but it is important to recognize that the product must have differed from the natural polymer in many respects.

For the next few years synthetic rubber research was minimal but two stimuli occurred about 1926. First, Staudinger was establishing the long chain nature of the rubber molecule and secondly the price of rubber was rising once again.

Whilst the peroxide-initiated emulsion polymerized polybutadiene had disappointing properties it was found in 1929 that copolymerization of butadiene with styrene and in 1930 with acrylonitrile led to the production of interesting materials. The butadiene-styrene rubber, Buna S, was potentially a general purpose rubber but at that time not competitive with natural rubber. On the other hand the butadiene-acrylonitrile rubber, Buna N, now commonly known as nitrile rubber had certain properties such as oil resistance not shown by natural rubber and commercial production was started about 1935①. Commercial production of butadiene-styrene rubber did not commence until 1937 and many things were to happen before it became the world's most used rubber.

Soon after this the Du Pont company commercially introduced what is still the world's most important special purpose rubber. This had its origin in work on acetylene chemistry started by Newland in 1906 and which led in the mid-1920s to the development of a process for making vinyl acetylene from acetylene. The addition of hydrogen chloride to the vinyl acetylene yielded 2-chloro-1,3-butadiene, chloroprene, which could be polymerized to form a rubber with good heat, oil and ozone resistance. Initially marketed as Duprene in 1932 the Du Pont company subsequently changed the name to Neoprene. Similar materials are now produced by other organizations in France, Germany and Japan.

At about the same time American chemists were also responsible for the development of another elastomer-butyl rubber. In 1930 the Standard Oil Company of New Jersey entered into an agreement with IG Farben in Germany to assist each other in developing chemical products and processes from petroleum. This led the IG company to disclose to the American company that isobutylene could be polymerized to a high molecular weight polymer by a strong Lewis acid catalyst such as boron trifluoride at about −75℃②. The management of Standard Oil was initially primarily interested in the use of the material as a fuel additive but two chemists within the company, R. M. Thomas and W. J. Sparks were intrigued by the unusual elastomeric properties of the material. This stimulated work within the American company which led to the production of copolymers containing small amounts of isoprene. The latter provided a few double bonds in an otherwise saturated hydrocarbon which enabled the polymer to be cross-linked using more-or-less conventional vulcanizing systems③. Commercial production of this material, known as butyl rubber commenced in 1942.

1. New words

distill[dis'til]*v.* 蒸馏，提取

fractionate['frækʃəneit]*v.* 分级，分馏

akin[ə'kin]*adj.* 酷似，类似的

turpentine['tə:pəntain]*n.* 松节油

stimuli['stimjulai]*n.* stimulus 的复数，刺激因素，促进因素

chloroprene['klɔ:rəpri:n]*n.* 氯丁二烯

ozone['əuzəun]*n.* 臭氧

intrigue[in'tri:g]*v.* 激起……的兴趣

2. Phrases and expressions

emulsion polymerized polybutadiene 乳化聚丁二烯

nitrile rubber 丁腈橡胶

special purpose rubber 特种橡胶

vinyl acetylene 乙烯基乙炔

2-chloro-1, 3-butadiene 2-氯-1, 3-丁二烯

butyl rubber 丁基橡胶

strong Leuis acid catalyst 强路易斯酸催化剂

boron trifluoride 三氟化硼

more-or-less 或多或少，大体上，大约，左右

3. Notes to the test

① On the other hand the butadiene-acrylonitrile rubber, Buna N, now commonly known as nitrile rubber had certain properties such as oil resistance not shown by natural rubber and commercial production was started about 1935. 另外，丁二烯-丙烯腈橡胶（Buna N，现在通常称作丁腈橡胶）具有某些天然橡胶所没有的特性，如耐油性能，其工业化生产约始于1935年。

② This led the IG company to disclose to the American company that isobutylene could be polymerized to a high molecular weight polymer by a strong Lewis acid catalyst such as boron trifluoride at about −75℃. 这使IG公司向美国公司透露了采用强路易斯酸催化剂（如三氟化硼）在−75℃下可把异丁烯聚合成高相对分子质量聚合物的秘密。

③ The latter provided a few double bonds in an otherwise saturated hydrocarbon which enabled the polymer to be cross-linked using more-or-less conventional vulcanizing systems. 异戊二烯能为饱和烃提供额外的双键，以便使该种聚合物可以用常规的硫化体系进行交联。

Lesson 33 Rubber Materials

The materials used in these categories all have finite duties to perform within a rubber formulation. Therefore, each will be discussed in the sequence already listed earlier in last two lessons

Rubbers (hydrocarbon) Reclaim rubber can play an important part in the function of general or specialised compounds especially as a softening agent with natural rubber. Before the War it was used as an extender or cheapening agent, but at the present time it is used in its own right to confer special processing characteristics to specific compounds, e. g. to reduce nerve during processing, to improve adhesion in certain rubber to metal bonds and also to achieve good tack in friction compounds for use in belts, hose, etc.

Vulcanizing Agents: Vulcanisation is the industry term used to describe the process whereby rubbers are reacted with chemicals, usually in the presence of heat, to convert the thermoplastic, uncured state, into the generally accepted 'rubbery' or 'elastic' state. Natural (NR), SBR, polyisoprene (IR), butyl (IIR) and nitrile (NBR) rubbers (just to name a few) all react with sulphur and sulphur bearing chemicals to achieve this. However, with neoprenes (CR) metallic oxides such as zinc oxide and magnesium oxide serve as the vulcanizing agents.

Activators: Zinc oxide, stearic (or other fatty acid) are generally found in all recipes, based upon NR, SBR, IR, IIR, where they are used as activating materials and produce a uniform rate and state of cure in the compound.

However, they also appear in CR compounds but for different reasons. In this case, the fatty acid helps as an anti-roll sticking additive, whereas the zinc oxide is the vulcanizing agent. As well as zinc oxide, the oxides of calcium, magnesium and lead (white, yellow and red respectively) may also be used, to confer special properties, such as increased water resistance in the case of the lead oxides.

Accelerators: the majority of rubber recipes in use, incorporate organic acceleration, but other rubbers especially CR types and especially the 'G' types, use inorganic oxides, such as zinc oxide, to achieve a state of vulcanisation. Others such as EPDM occasionally use organic peroxides but these are the exception rather than the rule and with full time everyday use, these exceptions will soon be memorised①. In the meantime, it is always advisable to check the literature. Indeed even when experienced it is good practice to double check, as errors can and do creep in on occasion.

Fillers and Extenders: As a generalisation fillers can be termed reinforcing and non-reinforcing, and also fall into black and non-black types.

Dilution with non-black materials, such as china clay, talc and/or whiting is also practised with general rubber, and many industrial goods. This is not only to keep the cost down, which is very important, but also as a device to 'smooth' out the compound to help its processability, especially in extending calendering operations. Practical experience soon indicates just what is necessary to achieve processability, at the right cost levels.

Processing Aids: Rubber chemicals within this grouping include peptizing agents, softeners (oils and waxes) and plasticizers. A processing aid plays one or more of the roles included in the following list:

(a) Speeds up the rate of polymer breakdown and also controls the degree of break down.

(b) Helps to disperse the other compounding ingredients especially blacks.

(c) Helps to reduce nerve within the compound, and also shrinkage during subsequent processing.

(d) Can impart building tack to the compound.

(e) Improved and more stable compound processing, especially in the compound preparation (blanking) and molding areas.

Antioxidants/Antiozonants: It is almost universally necessary to add antioxidants and/or antiozonants to any polymer to impart improved and satisfactory ageing properties in the cured compound.

The earlier antioxidants used to help natural rubber life prolongation were based on various aromatic amines and phenols, and even today such materials are still used[2]. It is usual to use quantities of the order of 1 part to 100 parts of polymer, but the amount obviously must and does depend on the service requirements of the product and also upon the actual basic polymer in the formulation. Some antioxidants are specific to the type of protection which they impart, and can be used to improve heat resistance, flex cracking, or improved resistance to weathering, which covers the attack of oxygen and ozone, plus ultraviolet light, temperature variations and moisture.

Typical Formulations: The following recipes are given as a guide only, but may be used as a basis from which the ultimate compound properties required may be derived[3].

Cables

Insulation recipe

NR	100	Stearic acid	1	MRX	10
Antioxidant	1	Zinc oxide	5	Talc	100
Whiting	50	MBTS	1	sulphur	3

1. New words

tack[tæk] n. 黏性
NR n. 天然橡胶
SBR n. 丁苯橡胶
nitrile['naitri:l] n. 丙烯腈,腈类
neoprene['niəupri:n] n. 氯丁胶
magnesium[mæg'ni:zjən] n. 镁
EPDM n. 三元乙丙橡胶
whiting['hwaitiŋ] n. 白粉,铅粉
nerve[nə:v] n. 回缩性
antioxidant['ænti'ɔksidənt] n. 防老剂
MRX n. 矿质橡胶
MBTS n. 促进剂 MBTS

2. Phrases and expressions

reclaim rubber 再生橡胶
softening agent 软化剂
magnesium oxide 氧化镁
peptizing agent 塑解剂
flex cracking 屈挠龟裂
insulation recipe 绝缘配方

3. Notes to the text

① Others such as EPDM occasionally use organic peroxides but these are the exception rather than the rule and with full time everyday use, these exceptions will soon be memorised. 其他如三元乙丙橡胶有时也使用有机过氧化物,但这些是例外的情况,而不是作为经常使用的一般规则,这些例外的情况会很快被人们记住。

② The earlier antioxidants used to help natural rubber life prolongation were based on various aromatic amines and phenols, and even today such materials are still used. 以往用来延长天然橡胶使用寿命的防老化剂是各种芳香胺与酚类,即使直到今天这些材料仍在继续使用。

③ The following recipes are given as a guide only, but may be used as a basis from which the ultimate compound properties required may be derived. 下面的配方仅给出了一般规则,但是可以它为基础从中推导出胶料所需要的主要性能。

4. Exercises

(1) What does the word "vulcanizing" mean in the third paragraph?

(2) According to paragraph 3, what are the vulcanizing agents of neoprenes?

(3) The five roles of processing aids in rubber processing are ________, ________, ________, ________, and ________.

(4) What are the aims to add antioxidants to any rubber?

(5) Translate the insulation recipe in the last paragraph into Chinese.

Reading Material

Rubber Mix and Compound Design

Armed now with these basic facts, the various ratios of the ingredients within the recipe may be considered[①]. However, before even attempting to formulate a compound, the following points must be very carefully considered:

(1) The finished article must meet the service conditions required and it is, therefore, important to obtain this information accurately, right at the start of the operations.

(2) It is also very important that the compound is designed at the right cost, otherwise the product will not sell. This, however, is a two way condition. It could well be that the service conditions are very severe, so a correct selling price is also required, commensurate with the conditions[②]. It is quite often easy to design if price does not matter. It is also easy to sell, by giving the material and goods away!

(3) Having designed the compound at the correct price level, the material must now be capable of being processed satisfactorily on available factory machines and equipment. This is essential for good factory practice to be maintained.

(4) Finally, any specified physical properties must be obtained consistently in the finished product, and allowances made for the lower levels of dispersion obtained in the factory (even under supervised conditions), when comparison is made with results from laboratory work. It is a good idea to allow for an approximately 10% fall-off in properties between the laboratory and factory mixed materials[③].

In order to meet the specified requirements it is of course essential to become familiar with all the basic properties of the various polymers currently available and to blend as appropriate. Quite often it is necessary to arrive at compromises of the above few points and if absolutely necessary to discuss the specification with the customer. Very often these specifications are not compiled by rubber chemists or technologists and unfortunately impossible and diametrically opposite properties are written in at one and the same time. If the correct approach is made and once mutual confidences are established, these difficulties can be ironed out with the customer.

It is a widely used practice to base and express all the various ingredient quantities as parts per hundred of rubber hydrocarbon. Thus, as a general guide the ultimately derived basic recipe, discussed previously, would be as follows:

Rubber	100	Sulphur	2.5 ~ 3.5
Activator	1 ~ 5	Accelerator	0.5 ~ 1.5
Filler	As required	Softener	5 ~ 10
Antioxidant	1 ~ 2		

It becomes immediately apparent that an infinite number of permutations becomes possible between the ratios and amounts of the various ingredients and, indeed, this is the case. The reason for this is that service conditions of most of the individual products are also infinite in possibilities, and a glance at the typical formulations shown later will give a guide and basis from which the exact conditions required can be derived④.

At this stage it is considered pertinent also to mention that the vast majority of suppliers of 'rubber chemicals' and synthetic rubbers, etc., to the trade, provide technical back-up to their materials in the form of literature. Subsequently they should be consulted for advice at an early stage if difficulties arise.

1. New words

commensurate[kə'menʃərit] *adj.* 相应的,相称的
compromise['kɔmprəmaiz] *n.* & *v.* 妥协,和解,兼顾
technologist[tek'nɔlədʒist] *n.* 工艺师
diametrically[ˌdaiə'metrikəli] *adv.* 完全地
mutual['mjuːtjuəl] *adj.* 相互的,共同的
consult[kən'sʌlt] *v.* 咨询

2. Phrases and expressions

finished article　成品
back-up　返回

3. Notes to the text

① Armed now with these basic facts, the various ratios of the ingredients within the recipe may be considered.　弄清这些基本内容以后,就可以考虑配方中各组分的不同配比了。

② This, however, is a two way condition. It could well be that the service conditions are very severe, so a correct selling price is also required, commensurate with the conditions.　但这有两种情况,如使用条件很苛刻,那就需要有相应的销售价。

③ Finally, any specified physical properties must be obtained consistently in the finished product, and allowances made for the lower levels of dispersion obtained in the factory (even under supervised conditions), when comparison is made with results from laboratory work. It is a good idea to allow for an approximately 10% fall-off in properties between the laboratory and factory mixed materials.　最后,务必使成品所规定的物理性能保持稳定。在工厂大生产中(即使在控制条件下),允许各项指标比实验室得出的结果略低一些。允许工厂大生产比试验室的混炼胶性能大约低10%是可行的。

④ The reason for this is that service conditions of most of the individual products are also infinite in possibilities, and a glance at the typical formulations shown later will give a guide and basis from which the exact conditions required can be derived.　其原因是大多数产品的使用条件也可能有各种各样的变化,所以最后所列的典型配方只是给出一个原则和基础,据此可以引申出所需要的正确条件。

Lesson 34 Synthetic Fibers(1)

Synthetic fibers can be produced from a wide variety of substances: polyesters, polyamides, vinyls, etc. The same material can often be used in molding as well as spinning processes. Some of the polymers, especially those which are suitable thermoplastics, are largely of petroleum origin; others, such as the polylactones, represent an attempt to develop new materials from renewable and biodegradable resources.

Industrial production aims for continuous, one-stage, energy-efficient processes which can be conducted at reasonable temperatures. Conventional catalysts which accomplish this goal at acceptable rates and good yields unfortunately have a deleterious effect on other properties such as color, thermal stability, and mechanical properties. It is generally true that the improvement in one property is accomplished at the expense of another[①]. Flame retardants may lead to discoloration and low tensile elongation at break point; carriers which improve affinity for acid or basic dyes may lead to inferior mechanical properties.

Each type of polymeric material has its own outstanding qualities and its own defects. Polyethylene terephthalate, the preferred commercial polyester, has superior mechanical properties and chemical resistance; however, its dimensional stability and transparency are poor because of low heat distortion temperatures and high rates of crystallization. Impact strength is also poor, making this polymer unsuitable for molded articles. Polycarbonates, on the other hand, have a high heat distortion temperature and superior transparency, but poor resistance to chemicals. Polyesters with improved tensile strength, flame resistance, and thermal stability, which still retain their strong abrasion resistance and have low enough melt viscosities for ease in processing, are eagerly sought[②].

Polyamides are outstanding in physical strength and toughness, and have superior mechanical, electrical, and shaping properties and good chemical stability. Atmospheric degradation and poor thermal stability at melt temperature are the problems with these resins, but degradation still leaves those qualities superior compared to competitive polymers.

Nylon fibers are stronger than any natural fibers, have an abrasion resistance four times that of wool, and are unaffected by dry cleaning solvents. But nylon does not always dye uniformly, and since the mechanism of its pyrolysis is not known, it is difficult to predict the success of flame retardants. Low moisture absorption is a further problem, leading to static cling and general lack of comfort.

Polyimides have exceptional thermal stability, but cannot be melt processed. Acrylics and modacrylics also must be processed by wet or dry spinning. This not only requires solubility in a convenient solvent, but leads to microscopic cavities in the resulting filament and an accompanying tendency to shrink in hot water③.

Copolymers, especially block copolymers, can provide the low processing temperature or other desirable attribute of one of the components coupled with some superior properties of the other.

The demand for flame-retardant fabrics has increased. Confusability is also an important safety factor, since it is possible to incur severe burns by contact with molten polymers. It is also desirable for smoldering textile goods to generate as little smoke or poisonous gas as possible.

There is an ongoing effort to duplicate, in the synthetics, some of the properties of natural fibers such as absorbency and water retention.

1. New words

polylactone[ˌpɔli'læktəun] *n.* 聚内酯

deleterious[deli'tiəriəs] *adj.* 有害的，有害杂质的

affinity[ə'finiti] *n.* 亲合力，亲和力

pyrolysis[pai'rɔlisis] *n.* 热解作用，高温分解

cling[kliŋ] *vi.* 黏着，依附于

modacrylic[ˌmɔdə'krilik] *n.* 改性聚丙烯腈

2. Phrases and expressions

aim for... 目标是

polyethylene terephthalate 聚对苯二甲酸乙二酯

3. Notes to the text

① It is generally true that the improvement in one property is accomplished at the expense of another. 通常总是改善了一种性能，却同时损失了另一种性能。

② Polyesters with improved tensile strength, flame resistance, and thermal stability, which still retain their strong abrasion resistance and have low enough melt viscosities for ease in processing, are eagerly sought. 人们渴望能找到抗张强度、阻燃性以及热稳定性提高了但仍能保持很强的耐磨性和易于加工的低熔体黏度聚酯。

③ This not only requires solubility in a convenient solvent, but leads to microscopic cavities in the resulting filament and an accompanying tendency to shrink in hot water. 这不仅需要溶解在一种合适的溶剂中，而且会在成晶纤维中造成微小的空洞，同时会在热水中发生收缩。

4. Exercises

(1) Which of the following is not a synthetic fiber?

a. polyester b. polyimide c. silk d. nylon

(2) Which of the following is not right according to the text?

a. Polyethylene terephthalate has its own outstanding qualities and its own defects.

b. Flame retardants may improve affinity for acid or basic dyes.

c. Polyamides are outstanding in physical strength and toughness, but they have their own defects.

d. Nylon fibers don't always dye uniformly.

(3) Which of the following is the outstanding quality of polyimide?

a. Superior abrasion resistance. b. Superior water retention.

c. Low melt viscosities. d. Exceptional thermal stability.

(4) What properties of natural fibers are expected to be duplicated in the synthetic fibers?

a. Abrasion resistance. b. Absorbency.

c. Electrical properties. d. Shaping properties.

Reading Material

Synthetic Fibers (2)

While many of the polymers used for synthetic fibers are identical to those in plastics, the two industries grew up separately, with completely different terminologies, testing procedures, and so on. Many of the requirements for fabrics are stated in nonquantitative terms such as "hand" and "drape" which are difficult to relate to normal physical property measurements[①], but which can be critical from the standpoint of consumer acceptance, and therefore the commercial success, of a fiber.

A fiber is often defined as an object with a length-to-diameter ratio of at least 100. Synthetic fibers are spun in the form of continuous filaments, but may be chopped to much shorter *staple*, which is then twisted into thread before weaving. Natural fibers, with the exception of silk, are initially in staple form. The thickness of a fiber is most commonly expressed in terms of denier, which is the weight in grams of a 9,000-m length of the fiber. Stresses and tensile strengths are reported in terms of tenacity, with units of grams/denier.

Fiber processing

The polymer molecules in synthetic fibers are only slightly oriented by flow as they emerge from the spinnerette. To develop the tensile strengths and moduli necessary for textile fibers, the fibers must be drawn (stretched) to orient the molecules along the fiber axis and develop high degrees of crystallinity. All successful fiber-forming polymers are crystallizable, and so from a molecular standpoint, the polymer must have polar groups, between which strong hydrogen bonding holds the chains in a crystal lattice (e. g, polyacrylonitrile, nylons), or be sufficiently regular to pack closely in a lattice held together by dispersion force (e. g., isotactic polypropylene).

It was pointed out that the cross section of fibers is determined by the cross section of the spinnerette holes and the nature of the spinning process. This plays an important role in establishing the properties of the fiber. For certain applications, the spun fibers are textured after spinning. Carpet fibers, for example, are often given a heat-set twist and/or are crimped by passing them through a pair of gear-like rollers.

Dyeing

The dyeing of fibers is a complex art in itself. A successful dye must either form strong secondary bonds to polar groups on the polar, or react to form covalent bonds with functional groups on the polymer[②]. Furthermore, since the fibers are dyed after spinning, the dye must penetrate the fiber, diffusing into it from the dye bath. The dye molecules can't penetrate the crystalline areas of the polymer, so it is mainly the amorphous regions that are dyed. This often conflicts with the requirement of high crystallinity[③]. The chains of polyacrylonitrile, for example, while possessing the necessary polar sites for dye attachment in abundance, are so strongly bound to each other that it's difficult for the dye to penetrate. For this reason, acrylic fibers usually contain minor amounts of plasticizing comonomers to enhance dye penetration. Nonpolar, nonreactive fibers such as polypropylene, on the other hand, have no sites to which the dye can bond even if it could penetrate. This was long a problem with polypropylene fibers, and was initially overcome by incorporating a finely divided solid pigment in the polymer before melt spinning. Copolymers of propylene and a monomer with dye-accepting sites are now available.

Additives and treatments

Static electricity can be a big problem with carpets. Many carpet fibers therefore incorporate an anti-static agent to bleed off static charge.

The same polar bonding sites that make fibers dyeable also make them stainable. In the past, the finished item, a carpet, for example, would be treated with an anti-staining agent such as a fluorocarbon telomer. Nowadays, these coatings are often applied to the fibers before weaving. Another approach is to make use of bicomponent fibers. For example, a core fiber of nylon 6/6 could be coated with a sheath of highly stain-resistant polypropylene.

Effects of heat and moisture

The polarity of the polymer also directly influences its degree of water absorption. Other things being equal, the more polar the polymer, the higher its equilibrium moisture content under any given conditions of humidity. As with dyes, however, moisture content is reduced by strong interchain bonding. The moisture content exerts a strong influence on the feel and comfort of fibers. Hydrophobic fibers tend to have a "clammy" feel in clothing, and can build up static electricity charges.

Perhaps the most important effect of moisture on polar polymers is as a plasticizer. Since fiber-forming polymers are linear, heat is also a plasticizer. This explains why suits wrinkle on hot, humid days, and why the wrinkles can be removed by steam pressing. "Wash-and-wear" and "permanent-press" fabrics are produced by operations that cross-link the fibers by reacting with functional groups on the chains, such as the hydroxyls on cellulose. The more hydrophobic polymers are inherently more wrinkle resistant because they are not plasticized by water. Wash-and-wear shirts, therefore, usually are made of blends of polyethylene terephthalate and cotton, about 65%/35%.

1. New words

terminology[ˌtəːmiˈnɔlədʒi] *n.* 专业术语
drape[dreip] *v.* 悬挂
spin[spin] *v.* 纺
filament[ˈfiləmənt] *n.* 细丝
staple[ˈsteipl] *n.* 纤维产品
weave[wiːv] *v.* 织,编织
polar[ˈpɔlə] *adj.* 极性的
polyacrylonitrile[ˈpɔliˌækriləuˈnaitril] *n.* 聚丙烯腈
isotactic[ˌaisəuˈtæktik] *adj.* 等规的
polypropylene[ˈpɔliˈprɔpilin] *n.* 聚丙烯
spinnerette[ˌspinəˈret] *n.* 纺丝头,喷丝头
texture[ˈtekstʃə] *n.* (织物)组织,构造; *vt.* 使具有某种结构(组织)
carpet[ˈkaːpit] *n.* 地毯
dye[dai] *v.* 染色
diffuse[diˈfjuːz] *v.* 扩散
acrylic[əˈkrilik] *adj.* 丙烯酸的
additive[ˈæditiv] *n.* 添加剂,助剂
stainable[ˈsteinəbl] *adj.* 可着色的
sheath[ʃiːθ] *n.* 鞘,套
moisture[ˈmɔisʃə] *n.* 水分
clammy[ˈklæmi] *adj.* 滑腻的,黏的
inherent[inˈhiərənt] *adj.* 固有的

2. Phrases and expressions

synthetic fiber 合成纤维
identical to 与……一致
twist into thread 捻成一根
natural fiber 天然纤维
degree of crystallinity 结晶度
polar group 极性基团
dispersion force 分散力
gear-like roller 齿轮状辊
static electricity 静电
bicomponent fiber 二组分纤维

3. Notes to the text

① Many of the requirements for fabrics are stated in nonquantitative terms such as "hand" and "drape" which are difficult to relate to normal physical property measurements. 许多对织物的要求用像"手感"和"垂度"等非定量的术语来说明,这些术语很难与通常的物理性能度量相关联。

② A successful dye must either form strong secondary bonds to polar groups on the polar, or react to form covalent bonds with functional groups on the polymer. 成功的染色必须与极性基团形成强的次价力键,或者通过反应与聚合物的官能团形成共价键。

③ The dye molecules can't penetrate the crystalline areas of the polymer, so it is mainly the amorphous regions that are dyed. This often conflicts with the requirement of high crystallinity. 染色分子不能穿透聚合物的结晶区域,所以主要是无定形区域被染色。这常常与高结晶度要求产生冲突。

Lesson 35 Adhesives

Adhesives have been a technologically important application of polymers for thousands of years. Many of the early natural adhesives are still used. These include starch and protein-based formulations such as hydrolyzed collagen from animal hides, hooves, and bones and casein from milk. As new adhesive formulations based on synthetic polymers (often the same polymers used in other applications) continue to be developed, the range of applications for adhesives has expanded dramatically.

An adhesive has been defined as a substance capable of holding materials (adherends) together by surface attachment. Adhesives offer a number of significant advantages as a means of bonding: (1) They are often the only practical means available, particularly in the case of small adherends. For example, it's hard to imagine welding abrasive grains to a paper backing to make sandpaper, or bolting the grains together to make a grinding wheel. (2) In the adhesive joining of large adherends, forces are fairly uniformly distributed over large areas of the adherend, resulting in low stresses, and holes (necessary for riveting or bolting), which invariably act as stress concentrators in the adherends, are eliminated, thus lowering the possibility of adherend failure①. (3) In addition to joining, adhesives may also act as seals against the penetration of fluids. In the case of corrosive fluids, this, coupled with the absence of holes, where corrosion usually gains an initial foothold, can minimize corrosion problems. (4) In terms of weight, it doesn't take much adhesive to join much larger adherends. Hence, it is not surprising that many of the newer high-performance adhesives were originally developed for aerospace applications. (5) Adhesive joining may offer economic advantages, often by reducing the hand labor necessary for other bonding techniques.

A detailed treatment of the science of adhesion is beyond the scope of this chapter. Nevertheless, some important generalizations will be drawn. Adhesion results from (1) mechanical bonding between the adhesive and adherend, and (2) chemical forces either primary covalent bonds or polar secondary forces between the two. The latter are thought to be the more important, and this explains in part why inert, nonpolar polymeric substrates such as polyethylene and polytetrafluoroethylene are very difficult to adhesive bond. They must first be chemically treated to introduce polar sites on the surface. To promote mechanical bonding, adherend surfaces are often roughened before joining, but this is sometimes counterproductive. It can trap air bubbles at the bottom of crevices which act as stress concentrators to promote failure in rigid adhesives②.

With good bonding between adhesive and adherend, joint failure is cohesive (the adhesive itself or the substrate fails). Where the adhesive is weaker than the substrate, to a good approximation, the properties of the adhesive polymer determine the properties of the adhesive joint; that is, the bond can be no stronger than the glue line③. Brittle polymers give brittle joints, polymers with high shear strengths give bonds of high shear strength, and heat-resistant polymers produce bonds with good heat resistance, and so on.

To form a successful joint, the adhesive must intimately contact the adherend surface. This requires first that it wet the surface. The subject of wetting is considered in detail in treatises on surface chemistry. In general, wetting is promoted by polar secondary forces between adhesive and substrate. This is another reason why low-polarity polymeric adherends such as polyethylene and polytetrafluoroethylene are difficult to bond with adhesives. To insure proper wetting and interfacial bonding, it is often necessary to clean the adherend surfaces carefully before joining. Good contact also requires a viscosity low enough under conditions of application to allow the adhesive to flow over the surface and into its nooks and crannies. Once contact has been established, the adhesive must harden to provide the necessary joint strength. There are five general categories of organic adhesive that accomplish these objectives in different ways.

1. New words

adhesive[əd'hi:siv] *n.* 黏合剂,胶黏剂; *adj.* 可黏着的,黏性的
starch[sta:tʃ] *n.* 淀粉
hide[haid] *n.* 兽皮,皮革
hoof[hu:f] *n.* (兽的)蹄,马蹄
casein['keisi:in] *n.* 干酪素,酪蛋白
adherend[əd'hiərənd] *n.* (使用黏合剂的)黏着面,被黏物(黏附体)
welding['weldiŋ] *n.* 焊接法,定位焊接
bolt[bəult] *vt.* & *vi.* 用螺栓拴上
rivet['rivit] *n.* 铆钉, *vt.* 铆接
corrosive[kə'rəusiv] *adj.* 腐蚀性的,侵蚀性的
inert[i'nə:t] *adj.* 惰性的
polytetrafluoroethylene [ˌpɔliˌtetrəˌfluərə'eθili:n] *n.* 聚四氟乙烯
counterproductive['kauntəprəˌdʌktiv] *adj.* 反生产的,使达不到预期目标的
crevice['krevis] *n.* 裂缝,缺口
cohesive[kəu'hi:siv] *adj.* 有结合力的,产生结合力的,产生内聚力的
wetting['wetiŋ] *n.* ;& *v.* 润湿
nook[nuk] *n.* 隐蔽处,死角
cranny['kræni] *n.* 裂缝,裂隙

2. Phrases and expressions

hydrolyzed collagen 水解胶原蛋白
surface attachment 表面吸附
abrasive grains 砂磨粒
grinding wheel 砂轮
stress concentrator 应力集中点
polar secondary force 极性次价力
shear strength 剪切强度

3. Notes to the text

① In the adhesive joining of large adherends, forces are fairly uniformly distributed over large areas of the adherend, resulting in low stresses, and holes(necessary for riveting or bolting), which invariably act as stress concentrators in the adherends, are eliminated, thus lowering the possibility of adherend failure. 粘接大的被黏物时,力总是大范围地并很均匀地分布在被黏物上,从而产生的应力较低,消除了在被黏物中总是充当应力集中点的孔洞(这些孔洞对铆接或螺栓连接是必要的),降低了粘接失败的可能性。

② It can trap air bubbles at the bottom of crevices which act as stress concentrators to promote failure in rigid adhesives. 它会在作为应力集中点的裂缝底部捕获气泡,导致使用刚性胶黏时的失效。

③ Where the adhesive is weaker than the substrate, to a good approximation, the properties of the adhesive polymer determine the properties of the adhesive joint; that is, the bond can be no stronger than the glue line. 一般说来,当黏合剂的黏合强度低于被黏合材料的强度时,胶黏聚合物的性能基本能够决定粘接处的性能,即黏合物的强度不可能比胶缝强。

4. Exercises

(1) How many kinds of adhesives have people used in their daily life? Which of them are being used most often?

(2) Take a look at the objects in your room, such as a mobile phone or a computer, how do adhesives hold the parts of them together?

(3) When choosing an adhesive for adherends,

a. what factors should be taken into account first?

b. how do you promote mechanical bonding?

c. what are the processes of adhesive bonding?

(4) Have you ever considered the environmental problem that goes with the application of an adhesive? How do you possibly solve or avoid this problem?

Reading Material

Five Organic Adhesives

1. Solvent-Based Adhesives. Here the adhesive polymer is made to flow by dissolving it in an appropriate solvent to form a cement. The adhesive hardens by evaporation of the solvent. Thus, the polymers used must be linear or branched to allow solution, and the joints formed will not be resistant to solvents of the type used initially to dissolve the polymer. To get a good bond, it helps if the solvent attacks the adherend also. In fact, solvent alone is often used to "solvent weld" polymers dissolving some of the adherend to form an adhesive on application.

One of the drawbacks to solvent-based adhesives based on rigid polymers is the shrinkage that results when the solvent evaporates[①]. This can set up stresses that weaken the joint. An example of this type of adhesive is the familiar model airplane cement, basically a cellulose nitrate solution, with perhaps some plasticizer. Rubber cements, of course, maintain their flexibility, but cannot support as great a stress. Commercial rubber cements are based on natural, SBR (poly(butadiene-co-styrene)), nitrile (poly(butadiene-co-acrylonitrile)), chloroprene (poly(2-chlorobutadiene)), and reclaimed (devulcanized) rubbers. Examples are household rubber cement and bond. Rubber cements may also incorporate a curing agent to crosslink the polymer after application and evaporation of the solvent. This greatly increases solvent resistance and strength.

2. Latex Adhesives. These materials are based on polymer latexes made by emulsion polymerization. They flow easily while the continuous water phase is present and dry by evaporation of the water, leaving behind a layer of polymer. In order that the polymer particles coalesce to form a continuous joint and be able to flow to contact the adherend surfaces, the polymers used must be above their glass transition temperature at use temperature. These requirements are similar to those for latex paints, so it is not surprising that some of the same polymers are used in both applications, for example, styrene-butadiene copolymers and polyvinyl acetate. Nitrile and neoprene rubbers are used for increased polarity. A familiar example of a latex adhesive is "white glue." basically a plasticized polyvinyl acetate latex. Latex adhesives are displacing solvent-based adhesives in many applications because of their reduced pollution and fire hazards. They are used extensively for bonding pile to backing in carpets.

3. Pressure-Sensitive Adhesives. These are really viscous polymer melts at room temperature, so the polymers used must be above their glass transitions. They are caused to flow and contact the adherends by applied pressure, and when the pressure is released, the viscosity is high enough to withstand the stresses produced by the adherends, which obviously cannot be very great[②]. The key property for a polymer used in this application is tack, which basically is a viscosity low enough to permit good surface contact, yet high enough to resist separation under stress, something on the order of 10^4-10^6 cP, although elasticity probably plays a role, also. Natural, SBR, and reclaimed rubbers are common in this application. The many varieties of pressure-sensitive tape are faced with this type of adhesive.

Contact cements are a variation in which the rubbery polymer is applied to each adherend surface in the form of a solution, or increasingly, a latex. Evaporation of the solvent or water leaves a polymer film with the tack necessary to grab and hold the adherends when they are pressed together.

4. Hot-Melt Adhesives. Thermoplastics often form good adhesives simply by being melted to cause flow and then solidifying on cooling after contacting the surfaces under moderate pressure. Polyamides and poly (ethylene-co-vinylacetate) are used frequently as hot-melt adhesives. Electric "glue guns" have been introduced to the consumer market which operate on this principle.

5. Reactive Adhesives. These compounds are either monomers or low molecular weight polymers which solidify by a polymerization and/or crosslinking reaction after application. They can develop tremendous bond strengths and have good solvent resistance and good (for polymers, anyhow) high-temperature properties. The most familiar example of reactive adhesives are the epoxies generally cured by multifunctional amines. Polyurethanes also make excellent reactive adhesives.

The α-alkyl cyanoacrylate "super glues" ("one drop holds 5000lbs") are now a familiar part of the consumer market. Originally, the monomers had extremely low viscosities, and so could crawl into narrow crevices and wet the adherend surfaces rapidly. On the other hand, they wouldn't fill gaps, and were absorbed into porous adherends, giving poor bonds. Newer versions are available with higher viscosities to overcome these drawbacks. Cyanoacrylates can polymerize in seconds by an anionic addition reaction believed initiated by hydroxyl ions from water adsorbed on the adherend surfaces[③]:

$$x\mathrm{H_2C{=}C(CN){-}C({=}O){-}OR} \xrightarrow{\mathrm{OH^-}} \mathrm{HO{+}C(H)(H){-}C(CN)(C({=}O)OR){+}_{x}H}$$

Unfortunately (or fortunately, if you stick your fingers together), being linear and polar, they have poor resistance to polar solvents (acetone is a good solvent), and they are subject to hydrolysis, and so have poor environmental stability.

Phenolic and other formaldehyde condensation polymers are also important reactive adhesives. Powdered phenolic resin is mixed with abrasive grams and the mixture is compression molded to form grinding wheels. A B-stage phenolic in a solvent is used to impregnate tissue paper. The solvent is evaporated, and the dry sheets are placed between layers of wood in a heated press, where the resin first melts and then cures, bonding the wood to form plywood. Similarly, sheets of paper impregnated with a B-stage melamine-formaldehyde resin are laminated and cured to form the familiar Formica counter tops.

Unlike the previous examples of reactive adhesives, the phenolics and other formaldehyde condensation polymers evolve water as they cure. If trapped in the joint, this can result in serious weakness, which limits their adhesive applications.

Note that all these examples of reactive adhesives are highly polar polymers. It is largely this polarity that accounts for their good bonding capabilities.

1. New words

evaporate[i'væpəreit]*v.* 蒸发,失去水分,消失

plasticizer['plæstisaizə]*n.* 增塑剂

flexibility[ˌfleksə'biliti]*n.* 灵活性,柔韧性,弯曲性,挠性

nitrile['naitrail]*n.* 丙烯腈

devulcanized[di:'vʌlkənaizd]*adj.* 脱硫的

bond[bɔnd]*n.* 合成树脂结合剂

latex['leiteks]*n.* 胶乳,乳胶

coalesce[kəuə'les]*vi.* 联合,合并

alkyl['ælkil]*n.* 烷(烃)基

cyanoacrylate[ˌsaiənəu'ækrileit]*n.* 氰基丙烯酸盐黏合剂,氰丙烯酸酯

porous['pɔ:rəs]*adj.* 可渗透的,多孔的,疏松的

anionic['ænaiənik]*adj.* 阴离子的,具有活性阴离子的

acetone['æsitəun]*n.* 丙酮

hydrolysis[hai'drɔlisis]*n.* 水解,水解作用

formaldehyde[fɔ:'mældiˌhaid]*n.* 甲醛,蚁醛

evaporated[i'væpəreitid]*adj.* 浓缩的,脱水的,蒸发干燥的

plywood['plaiwud]*n.* 夹板,合板,胶合板

melamine['meləmi(:)n]*n.* 三聚氰(酰)胺,密胺

laminate['læmineit]*v.* 切成薄板(片);*n.* 层压板,层压材料

2. Phrases and expressions

reclaimed rubber　再生橡胶

rubber cement　橡胶胶水

solvent resistance　耐溶剂性

curing agent　固化剂,硫化剂

latex paint　乳胶涂料

polyvinyl acetate　聚乙烯乙酸酯

neoprene rubber　氯丁橡胶

pressure-sensitive tape　压敏胶黏带

contact cement　接触胶合剂

glue gun　黏合剂用喷枪,喷胶器,热融枪

condensation polymer　缩(合)聚(合)物

compression mold　压模

tissue paper　棉纸,薄纸

3. Notes to the text

①One of the drawbacks to solvent-based adhesives based on rigid polymers is the shrinkage that results when the solvent evaporates.　刚性聚合物的溶剂型胶黏剂的缺点之一是溶剂蒸发可产生收缩。

②They are caused to flow and contact the adherends by applied pressure, and when the pressure is released, the viscosity is high enough to withstand the stresses produced by the adherends, which obviously cannot be very great.　通过施加压力使得它们流过并接触被黏物,压力一旦解除,其黏度就足以承受被黏物产生的显然不会太大的应力。

③Cyanoacrylates can polymerize in seconds by an anionic addition reaction believed initiated by hydroxyl ions from water adsorbed on the adherend surfaces;　氰基丙烯酸盐黏合剂可通过一个阴离子加成反应在很短的时间内聚合,该反应被认为是由被黏物表面吸收的水分中的羟离子引发的。

PART 7 POLYMER MOLDING AND PROCESSING

Lesson 36 Plastics Molding and Processing

The resin manufacturing industries are those concerned with converting raw materials into chemical compounds which are further processed into the various plastics resins. These resins are produced in many forms; as granules, powders, liquids or pastes, ready to be sold to companies which will process them into finished products. The majority of these industries are made up of chemical companies, who have the scientists and equipment to produce plastics, and petroleum companies, who have vast raw materials and chemical laboratory experience (Figures 36. 1 and 36. 2). Some companies purchase chemicals and produce resins. Others manufacture resins and produce plastics in sheet, rod, tubing and other standard shapes.

Figure 36. 1 A petrochemical facility, this plant converts liquefied petroleum into gasses which are chemically processed into polyethylene and polypropylene

Companies that make plastic products may be divided into two groups, those that do molding and those that do fabricating. Companies

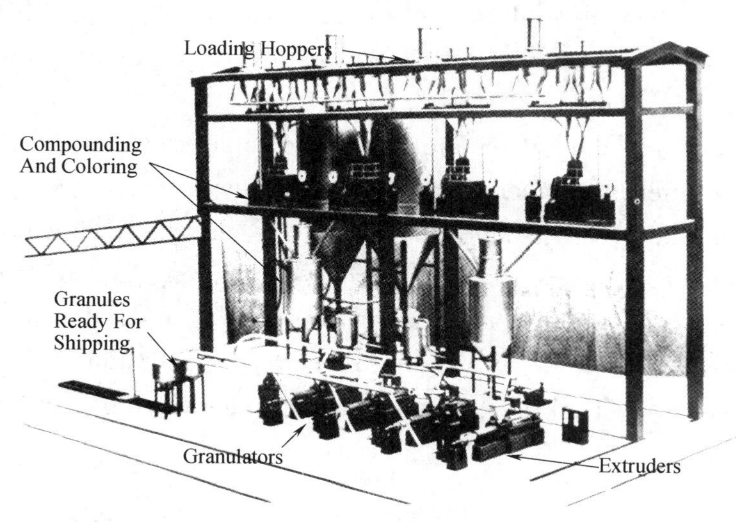

Figure 36. 2 Scale model of an automated mixing, compounding, palletizing plant. In a full-size plant the plastic is converted into granular form to be shipped to the processor

that usually do molding purchase plastics resins from a resin manufacturer and convert these materials into products by molding processes. Plastics resins with a wide range of colors and properties are available. Some companies take orders for large quantities of a single item within the range of their equipment. Other companies operate as job shops and produce many different low volume items. In some instances companies not related to the plastics industry buy molding equipment to make products for themselves. A good example is a dairy that blow molds milk bottles in their own plant(Figure 36. 3).

Fabricating companies differ from those that do molding in that they begin with standard sheet and stock plastics materials rather than with granules or powders[①]. Fabricators use many production techniques in manufacturing products(Figure 36. 4).

Figure 36. 3 In plant molding of ready-to-fill gallon milk bottles

Figure 36. 4 This display sign, which is 16ft. in diameter, has been thermoformed from plastic sheet, painted on the inside and cemented together

A number of companies are in the business of supplying materials and services to decorate of produce special finishes for plastics products. Some of this work is done by molders and fabricators themselves while other companies provide this service. Since most plastic products require some type of decorating, there are many companies closely related to the plastics industry that work in this area.

As with other materials, there are certain advantages and disadvantages of using plastics. Plastics differ from other materials in that they provide a combination of properties rather than extremes of single properties. This characteristic contributes much to their widespread usage. Plastics can be flexible, light, strong, and transparent all at the same time. These combination properties make plastics ideal for many applications where other materials with one outstanding property are not suitable. The following figure illustrates some major advantages and disadvantages of plastics materials.

1. New words

petrochemical[ˌpetrəu'kemikəl] *n.* 石化产品; *adj.* 石化的

compounding [ˌkɔmpaundiŋ] *n.* 组合，混合

granule['grænju:l] n. 颗粒
paste[peist] n. 糊(塑料)
fabricating['fæbrikeitiŋ] n. 二次加工
transparent[træns'pɛərənt] adj. 透明的
stock[stɔk] n. 毛坯,坯料
deterioration[diˌtiəriə'reiʃən] n. 恶化

2. Phrases and expressions

be concerned with 包括,与……有关
be divided into 被分成
ready-to-fill adj. 现成的,做好的
within the range of 在……范围内
job shop 加工车间
in the business of 从事
combination of properties 综合性能
make sth. ideal for 使……适合于

3. Notes to the text

① Fabricating companies differ from those that do molding in that they begin with standard sheet and stock plastics materials rather than with granules or powders. 与那些模塑公司不同的是,二次成型公司是用标准片材或坯料成型制品,而不是用粒料或粉料直接模塑制品。

4. Exercises

(1) Plastics provide a ________ of properties rather than ________ properties.

(2) In terms of properties, plastics can be flexible ________, ________, and ________ all at the same time.

(3) Companies that make plastic products may be divided into two groups, what are they?

(4) In what four forms do resin manufacturers make plastics ready for molding?

(5) Five major advantages of plastics are ________.

(6) Five major disadvantages of plastics are ________.

(7) List the different forms in which plastics resins are produced and try to obtain samples from manufacturers.

Reading Material

Rubber Product Manufacturing Systems

The manufacturing technology of the rubber industry is based on a number of key processes and operations, some of which have undergone a continuous evolution since the founding of the first rubber companies. These processes and operations form the building blocks from which a complex and diverse range of manufacturing systems are assembled. For conventional vulcanized rubber products, on which this book concentrates, a sequence of processes and operations is necessary to complete manufacture. It is possible to draw complex generalized flowcharts for the manufacture of typical products, but in the majority of cases these flowcharts can be reduced to three fundamental stages:

Mix→Shape→Vulcanize

The techniques used for each of these three stages will exert a substantial influence on productivity and on the quality of the finished products, both singly and in combination with each other. For example, it is well known that the quality and uniformity of mixing exerts a profound influence on the performance of downstream processes. Consequently, it is necessary to deal with rubber product manufacturing systems at three levels:

(1) Processing behavior of raw elastomers and rubber compounds.

(2) Unit processes.

(3) Manufacturing systems.

Determination of processing behavior is essential for the effective setting up and operation of unit processes, which include familiar operations such as mixing, extrusion, calendering, and molding. The unit processes then have to be assembled into a viable manufacturing system, which requires both organizational and technical skills.

In recent years the introduction of rubber compounds with widely differing processing characteristics, coupled with the need to achieve greater precision in manufacture to satisfy the increasingly stringent demands of customers, has stimulated a much greater interest in manufacturing methods as a whole①. The changes thus initiated have now been substantially accelerated by the advent of the microprocessor, with all its attendant implications for automation, process monitoring, and production organization.

Computer methods also have a substantial role to play in the support of manufacture. The increasing range of options for process operating sequences and conditions, with the accompanying potential for high productivity and quality, require improvements in the methods used for the selection of those options which give optimum performance for specific products②. Efficient and practical methods of determining process capability are now available which rely entirely on computer analysis. Similarly, die and mold design methods are substantially improved by adopting a computer-aided approach.

Integrating the introduction of computer methods into a company at a number of different levels, with all the attendant considerations of their influence on operating and organizational methods, is a complex under-taking. Fortunately, the systems approach, which is well-established in the production engineering and management fields, provides the concepts and techniques necessary to facilitate this integration③.

The systems approach to manufacturing is based on making a clear distinction between analysis and synthesis. Analysis is concerned with the identification, detailed study, and improvement of the fundamental elements of manufacturing systems. A system can be identified as being any activity that has well-defined inputs and outputs, which indicated that the scale of analysis can be established at many different levels. For example, an extrusion line can be treated as a complete system which has well-defined inputs and outputs and a number of interdependent elements (Figure 36. 5).

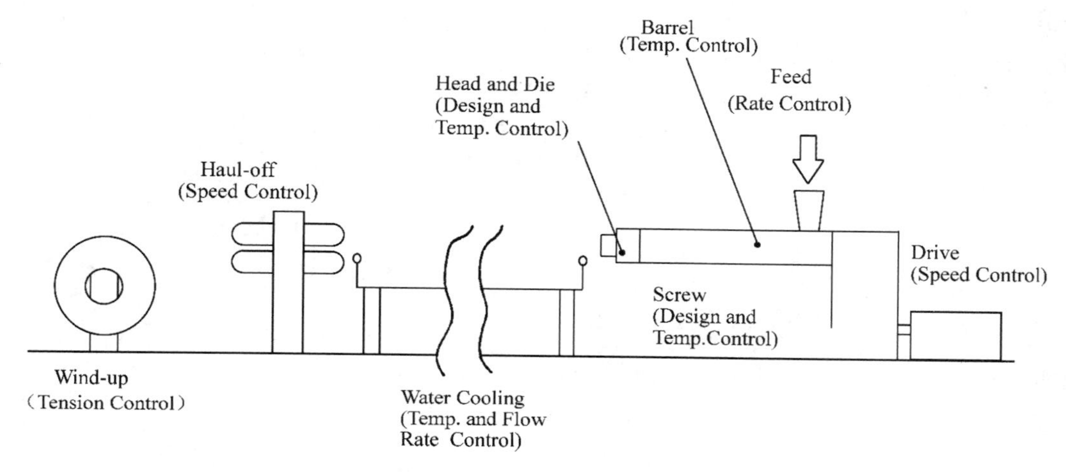

Figure 36. 5 Elements of an extruder line which interact to determine its overall performance

1. New words

evolution[evəˈluːʃən] *n.* 改进,进展

integration[ˌintiˈgreiʃən] *n.* 集成

element[ˈelimənt] *n.* 要素,元素,成分,元件,自然环境

screw[skruː] *n.* 螺杆,螺丝钉; *vt.* 调节,旋; *vi.* 转动,旋,拧

barrel[ˈbærəl] *n.* 料筒,桶; *vt.* 装入桶内

2. Phrases and expressions

generalized flowcharts 工艺流程图

as a whole 整个

computer-aided approach 计算机辅助方法

wind-up 收卷,收卷机,结局,结尾

haul-off(= take off) 引出,引出装置

tension control 张力控制

3. Notes to the text

① In recent years the introduction of rubber compounds with widely differing processing characteristics, coupled with the need to achieve greater precision in manufacture to satisfy the increasingly stringent demands of customers, has stimulated a much greater interest in manufacturing methods as a whole. 近年来,由于采用了加工特性条件很不相同的各种胶料,以及为满足用户对产品质量日益严格的要求而必须要达到加工的更高精度,激发了对改进整个加工方法的更大兴趣。

② The increasing range of options for process operating sequences and conditions, require improvements in the methods used for the selection of those options which give optimum performance for specific products.　加工操作顺序与条件可选范围的不断增加要求改进选择这些选项的方法,使具体的产品具有最佳性能。

③ Fortunately, the systems approach, which is well-established in the production engineering and management fields, provides the concepts and techniques necessary to facilitate this integration.　幸运的是,在生产工程和管理领域已经确立的系统方法可提供有助于这种集成综合所必需的概念和技术。

Lesson 37 Injection Molding

Many different processes are used to transform plastic granules, powders, and liquids into final products. The plastic material is in moldable form, and is adaptable to various forming methods. In most cases thermoplastic materials are suitable for certain processes while thermosetting materials require other methods of forming. This is recognized by the fact that thermoplastics are usually heated to a soft state and then reshaped before cooling. Thermosets, on the other hand have not yet been polymerized before processing, and the chemical reaction takes place during the process, usually through heat, a catalyst, or pressure. It is important to remember this concept while studying the plastics manufacturing processes and the polymers used.

Injection molding is by far the most widely used process of forming thermoplastic materials. It is also one of the oldest. The basic process involves six major steps in the molding cycle:

(1) The hopper is loaded with granular plastic materials.

(2) Heat is applied to the plastic until it becomes soft enough to flow.

(3) The softened plastic is forced through a nozzle into the mold cavity.

(4) Then cool, the halves of the mold are separated.

(5) The part is ejected from the mold.

(6) Gates connecting the product to the runner system are removed.

Variations of this basic molding process are involved in all injection molding. Figure 37. 1 illustrates some typical products produced by injection molding.

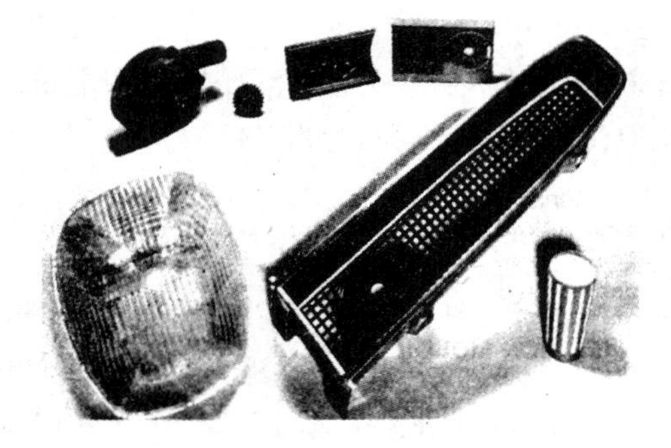

Figure 37. 1 Various types and sizes of injection molded parts include a highway light lens, industrial pump housing, gear, furniture part, radio housing, automobile grille, and a drinking cup

Molding Process: Plastic granules are fed into the hopper and through an opening in the injection cylinder where they are carried forward by the rotating screw①. The rotation of the screw forces the granules under high pressure against the heated walls of the cylinder causing them to melt. As the pressure builds up, the rotating screw is forced backward until enough plastic has accumulated to make the shot. At this point the screw is hydraulically forced forward, injecting the plastic through the nozzle, on through the sprue and runners, and into the cavities of the closed mold②.

Pressure is held briefly in order for the plastic to set, after which the screw retracts, releasing the pressure. Water cooled molds cause the plastic to cool quickly. The mold is opened and the part is ejected from the movable half of the mold, usually by air pressure or spring loaded ejector pins. The mold is then closed to begin another cycle.

1. New words

hopper['hɔpə] *n.* 料斗
nozzle['nɔzl] *n.* 喷嘴
vertical['və:tikəl] *adj.* 立式的,竖式的
cylinder['stilində] *n.* 料筒
gate[geit] *n.* 浇口
hydraulic[hai'drə:lik] *adj.* 液压的
runner['rʌnə] *n.* 流道
opening['əupniŋ] *n.* 加料口
accumulate[ə'kju:mjuleit] *vi.* & *vt.* 聚积 积累
shot[ʃɔt] *n.* 注射,注射量
hydraulically[hai'drə:likli] *adv.* 液压地
sprue[spru:] *n.* 注道

2. Phrases and expressions

in moldable form　处于可模塑的形式
be recognized by　由……来识别
molding cycle　模塑周期
be loaded with　装(载)着……
be ejected from　从……中取出
closed mold　合模
spring loaded　弹簧加载
ejector pins　顶杆

3. Notes to the text

① Plastic granules are fed into the hopper and through an opening in the injection cylinder where they are carried forward by the rotating screw.　塑料粒料被送进料斗后,通过料筒的加料口进入料筒,然后即被旋转的螺杆向前输送。

② At this point the screw is hydraulically forced forward, injecting the plastic through the nozzle, on through the sprue and runners, and into the cavities of the closed mold.　此时在螺杆通过液压方式向前推进的时候就将已经熔融的物料通过喷嘴、主流道、分流道注入闭合的模具型腔中。

4. Exercises

(1) The six major steps in the injection molding cycle are ________.

(2) Forming thermoplastic materials by injection molding is ________.

a. one of the most difficult methods

b. one of the most widely used methods

c. one of the newest methods

d. none of the above

(3) Parts are ejected from the mold by ________ or ________.

(4) Why are thermosetting resins most useful in standard injection molding?

Reading Material

Injection Molding Machines and Molds

Injection molding machines are manufactured in many sizes. These are rated according to size by the amount of material which can be injected in one cycle, which ranges from a fraction of an ounce in the small laboratory models to many pounds in large production equipment. Laboratory models are used in the research and development of new polymers and molding techniques.

There are two basic units to an injection molding machine: one for injecting the heated plastic and one for opening and closing the mold①. The first unit includes a feed hopper, heated injection cylinder, and an injection plunger or screw system. The second unit comprises a hydraulic operated moving platen and a stationary platen on which the halves of the mold are mounted (Figure 37. 2). Injection molding machines are also available in vertical models.

Figure 37. 2 The molding unit of an injection molding machine. Location on the movable platen is for mounting the mold half

There are many variations in injection molding machine design, however, the basic machines are of either the screw-ram, or plunger type. The main difference between these types is the method in which the plastic material is delivered from the hopper to the nozzle of the machine. Machines of the reciprocating-screw type are used more because of faster cycles, lower melting temperatures, and better mixing of the material (Figure 37. 3).

Molds used in injection molding consist of two halves; one stationary and one movable. The stationary half is fastened directly to the stationary platen and is in direct contact with the nozzle of the injection unit during operation. The movable half of the mold is secured to the movable platen and usually contains the ejector mechanism. There are many possible mold designs, including multiple piece molds for complicated parts (Fig-

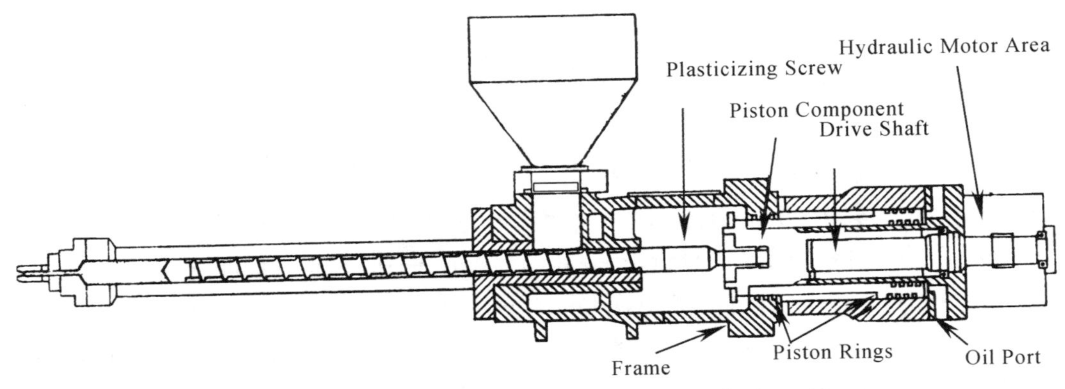

Figure 37. 3 A reciprocating screw machine on the back position. The screw is ready to be forced forward to make the injection shot

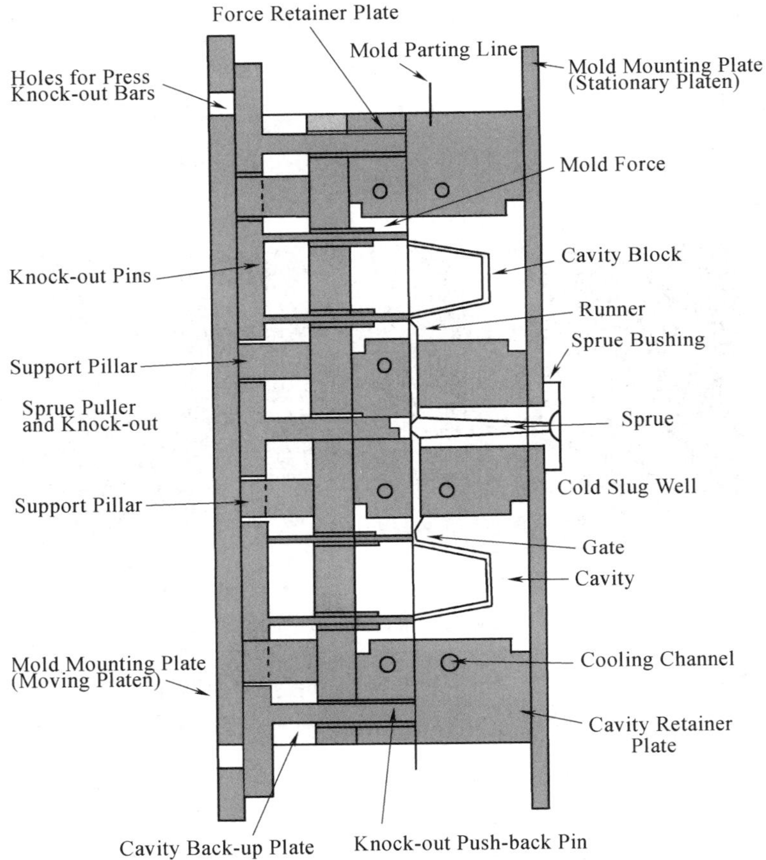

Figure 37. 4 Two-cavity injection mold

ure 37. 4). On production injection molding equipment many articles may be shot at the same time by the use of multiple cavity molds. The use of a balanced runner system carries the plastic from the sprue to each individual cavity. At this point the material passes through a gate into the cavity. The gate is a restriction, smaller than the runner, to provide for even filling of the mold cavity and to allow the products to be easily removed from the runner system. With most injection molding system, the articles can be snapped away from the runner or sprue without additional trimming. Products that have been injection molded can usually be identified by finding where the gate was broken off. The gate will usually be located at the edge or parting line of an object or in the center of cylindrical product②.

Injection molding is especially suitable for high production runs. Molds are expensive, as are the machines. Yet, once the product has been designed, molds made, and production started, articles can be produced in quantity at low cost. Most machines produce several thousand products an hour. Virtually all thermoplastics can be injection molded through variations in mold and machine design.

1. New words

port[pɔ:t] *n.* 端口,港口
plunger['plʌndʒə] *n.* 阳模,柱塞
mount[maunt] *vt.* 固定,安装
article['a:tikl] *n.* 制品
snap[snæp] *vt.* 折断,咬断,按扣
identify[ai'dentifai] *vt.* 鉴别

2. Phrases and expressions

injection molding machine 注塑机
be rated according to... 以……定等级或标定
one cycle (注射的)一个周期
a fraction of ……的几分之一
feed hopper 加料斗
moving platen 动模板
stationary platen 定模板
screw-ram 螺杆-射料杆式
reciprocating screw 往复式螺杆
be fastened to 固定在……上
ejector mechanism 顶出机构
multiple piece molds 多腔模具
piston ring 活塞环
two-cavity injection mold 双腔注模
balanced runner 平衡式流道
pass through 流过,通过
mold cavity 模具型腔
be broken off 脱落,折断
parting line 分模线,合模线
retainer plate 垫板
knock-out bars 顶出杆
support pillar 支承柱
mounting plate 固定板
back-up plate 支承板
cavity block 凹模
sprue bushing 浇口套
cold slug well 冷料井
cooling channel 冷却水通道
push-back pin 回程杆

3. Notes to the text

① ... one for injecting the heated plastic and one for opening and closing the mold. ……注射塑化系统和锁模系统。

② The gate will usually be located at the edge or parting line of an object or in the center of cylindrical product. 浇口通常设定在制品合模线的边缘或圆筒形制品的中心位置处。

Lesson 38 Extrusion

Thermoplastic items with a uniform cross section are formed by extrusion. This includes many familiar items such as pipe, hose and tubing, gaskets, wire and cable insulation, sheeting, window-frame moldings, and house siding. Molding powder is conveyed down an electrically or oil-heated barrel by a rotating screw. It melts as it proceeds down the barrel and is forced through a die which gives it its final shape (Figure 38. 1). Vented extruders incorporate a section in which a vacuum is applied to the melt to remove volatiles such as traces of unreacted monomer, moisture, solvent from the polymerization process, or degradation products.

The design of extruder screws is an interesting and complex technical problem and has received considerable study. Screws are optimized for the particular polymer being extruded. Basically, a screw consists of three sections: melting, compression, and metering. The function of the melting section is to convey the solid pellets forward from the hopper and convert them into molten polymer. Its analysis involves a combination of fluid and solid mechanics and heat transfer. The compression section, in which the depth of the screw flight decreases, is designed to compact and mix the molten polymer to provide a more-or-less homogeneous melt to the metering section, the function of which is to pump the molten polymer out through the die①.

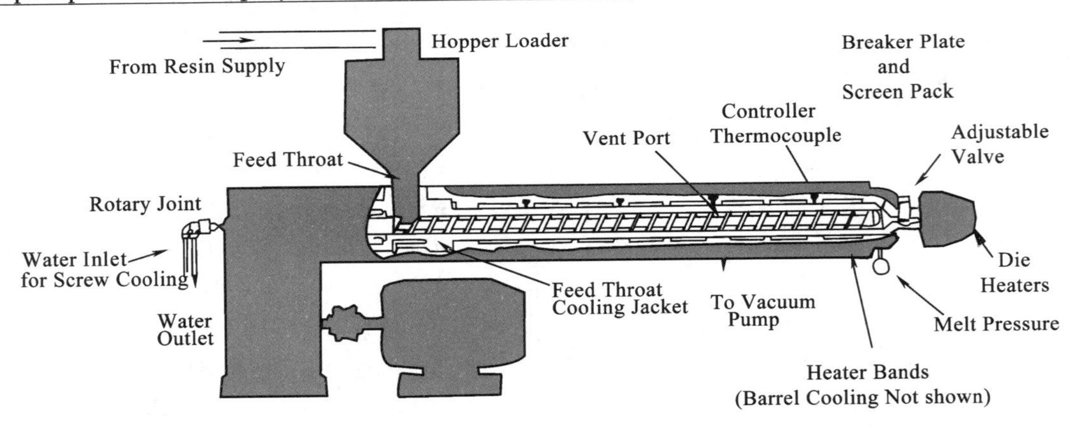

Figure 38. 1 Vented extruder

This last section is well understood. Analysis of the metering section is an interesting application of the rheological principles discussed previously, as is much of die design. The determination of the die cross section needed to produce a desired product cross section (other than circular) is still pretty much a trial-and-error process, however. Viscoelastic polymer melts swell upon emerging from the die (recovering stored elastic energy), and the degree of die swell cannot be reliably predicted②.

In addition to die swell, as the extrusion rate is increased, the extrudate begins to exhibit roughness, and then an irregular, severely distorted profile. This phenomenon is known as melt fracture. It is generally attributed to melt elasticity, but there is currently no way of quantitatively predicting its onset or severity. It can be minimized by increasing die length, smoothly tapering the entrance to the die, and raising the die temperature.

Extruders are normally specified by screw diameter and length-to-diameter ratio. Diameters range from 1 in. (2.5 cm) in laboratory or small production machines to 1 ft (30 cm) for machines used in the final pelletizing step of production operations. Typical *L/D* ratios seem to grow each year or so, with values now in the 20/1 to 36/1 range.

Single-screw extruders depend for their pumping ability on the drag flow of material between the rotating screw and stationary barrel. As a result, they are not positive-displacement pumps and tend to give a rather broad residence time distribution. Moreover, they are not particularly good mixing devices. Counter-rotating twin-screw extruders are true positive-displacement pumps, capable of generating the high pressures needed in certain profile extrusion applications. Co-rotating twin-screw extruders, though not positive displacement pumps, with proper screw design can give excellent mixing and a narrow residence time distribution (subjecting all the material to essentially the same shear and temperature history). They are, therefore, used extensively in polymer compounding (mixing) operations, and to a certain extent as continuous polymerization reactors.

In steady-state extruder operation, most or all of the energy needed to plasticize the polymer is supplied by the drive motor through viscous energy dissipation③. The heaters on extruders are needed mainly for startup and because enough heat can't always get from where it's generated (the compression and metering zones) to where it's needed (the melting zone). In fact, cooling through the barrel walls and/or screw center is sometimes necessary.

1. New words

hose[həuz] *n.* 水管,橡皮软管,胶管;*vt.* 用软管浇(冲洗)

tubing['tju:biŋ] *n.* 装管,配管,管道系统

gasket['gæskit] *n.* 衬垫,垫圈,垫片

incorporate[in'kɔ:pəreit] *vi.* 合并,合并的;*adj.* 具体化的

vacuum['vækjuəm] *n.* 真空,空间;*vt.* 用真空吸尘器清扫(某物)

volatiles['vɔlətail] *n.* 挥发组分,挥发物,挥发性物质

compression[kəm'preʃən] *n.* 挤压,压缩

metering['mi:təriŋ] *n.* 测量(法),计量,测定

pellet['pelit] *n.* 粒料,切粒

pump[pʌmp] *n.* 抽水机,泵;*vi.* 抽水,打气

die[dai]*n.* 模头,口模
viscoelastic[ˌviskəui'læstik]*adj.* 黏弹性的,黏弹性,黏滞弹性的
extrudate[iks'truːdeit]*n.* 挤出物,挤出料
distorted[dis'tɔːtid]*adj.* 扭歪的,受到曲解的
tapering['teipəriŋ]*adj.* 尖端细的,锥形的
palletizing['pelitaiziŋ]*n.* 微粒化
plasticize['plæstisaiz]*vt.* 增塑(塑炼),塑化

2. Phrases and expressions

cable insulation 电缆绝缘层,电缆绝缘料
vented extruders 排气式挤出机
extruder screws 挤出机螺杆
screw flight 螺杆的螺纹
cross section 横截面
trial-and-error 反复实验,尝试法,逐步逼近法,试错法
emerge from 从……露出,从……浮现,来自,产生于
extrusion rate 挤出速率
melt fracture 熔体破裂
screw diameter 螺杆直径
length-to-diameter ratio 长径比
drag flow 黏性流,阻曳流
residence time distribution 停留(阻滞)时间分布
counter-rotating twin-screw extruder 反向旋转双螺杆挤出机
energy dissipation 能量损耗
screen pack (挤出机的)滤网叠

3. Notes to the text

① The compression section, in which the depth of the screw flight decreases, is designed to compact and mix the molten polymer to provide a more-or-less homogeneous melt to the metering section, the function of which is to pump the molten polymer out through the die. 螺杆螺纹深度减少的压缩段是用来压紧并混合熔融高聚物,为计量段提供几乎均质的熔体,而计量段的作用是泵出熔融高聚物并将其从口模挤出。

② Viscoelastic polymer melts swell upon emerging from the die (recovering stored elastic energy), and the degree of die swell cannot be reliably predicted. 黏弹性高聚物熔体离开口模时会膨胀(恢复储存的弹性能),并且不能可靠地预测出离模膨胀的程度。

③ In steady-state extruder operation, most or all of the energy needed to plasticize the polymer is supplied by the drive motor through viscous energy dissipation. 在挤出机稳态运行时,大部分或者说全部用来塑化高聚物的能量是由驱动马达通过黏性能量损耗提供的。

4. Exercises

(1) Have you ever seen or operated extruders? How do they work?

(2) What are the functions of melting, compression and metering sections?

(3) What are the common extrusion defects? How to minimize their onset and severity to the lowest?

(4) What does length-to-diameter ratio mean to an extruder? How do you choose the right *L/D* ratio for a specific material?

(5) Translate the following sentence into Chinese:

The heaters on extruders are needed mainly for startup and because enough heat can't always get from where it's generated (the compression and metering zones) to where it's needed (the melting zone).

Reading Material

Extrusion and Flat Sheet Extrusion

In the extrusion process, thermoplastic resin is fed into a heated cylinder. The softened plastic is forced by a rotating screw, or plunger, through openings in accurately machined dies to form continuous shapes.

Extrusion is used to make three main types of products:

(1) Standard profile shapes such as rod, pipe, sheet, and irregular cross sections.

(2) Extrusions around wire and cable as a protective coating.

(3) Film to be used alone and as coatings for paper, cloth, and other surfaces.

Each of these types can be made in varying sizes depending upon the size of the machine. Extruders are designated as to size by the diameter of the rotating screw①. They range from large industrial extruders with screw diameters up to 6 or 8 in., down to the laboratory model with a 3/4 in. diameter screw. Extrusion is used only for the processing of thermoplastic resins.

Plastic granules are fed into the hopper of the extruder, picked up by the rotating screw within the hard liner of the extruder cylinder (Figure 38.2), and are forced forward. As the material moves along the cylinder, it is heated to a soft state and thoroughly mixed. The heat for softening the plastic comes from two sources; the external heater bands and the friction of the material against the rotating screw.

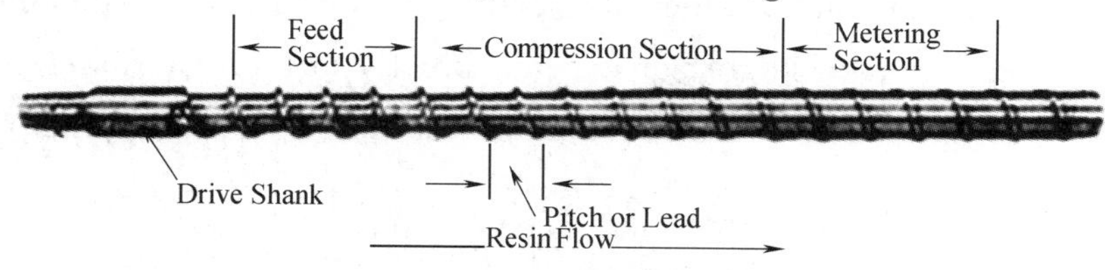

Figure 38.2 The parts of a 3 1/2 in. extruder screw

During the heating and compressing period the plastic must be transformed into a complete homogeneous (consistent) mix. This eliminates the possibility of wavy surfaces and a nonuniform cross section of the final product. It also provides for uniform color when the extruder is being used for color blending.

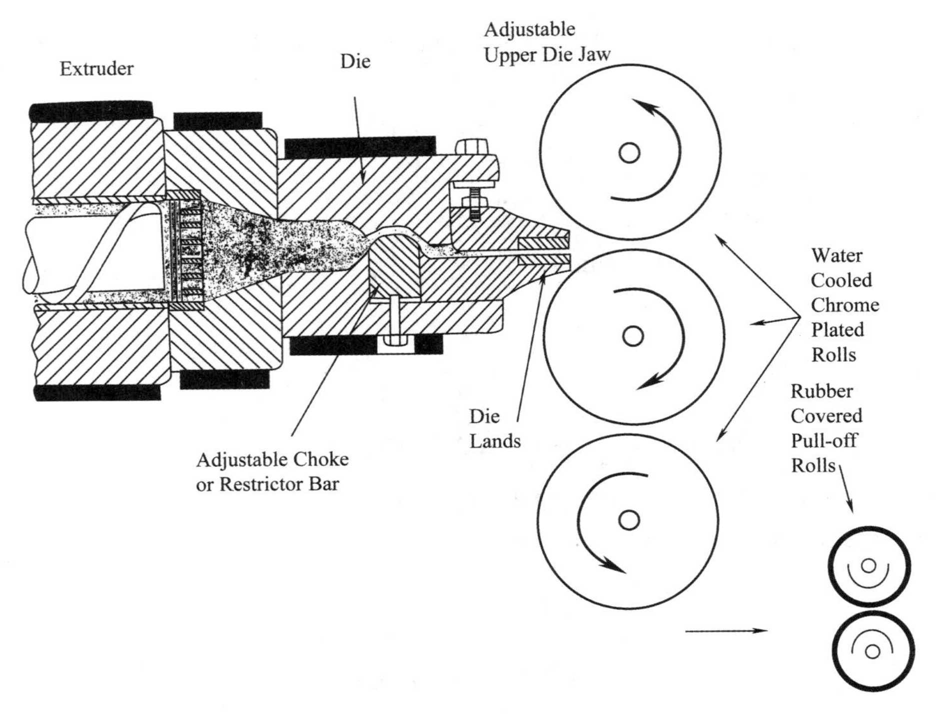

Figure 38.3 Cross section of a die for sheet extrusion and part of the take-off unit

The melted plastic finally passes through the screen pack, which removes dirt, and to the die at very high pressure②. From the die, the extruded profile passes through a cooling and take-off system. The final extrudate is wound onto coils or cut to specified lengths.

Film material over 0.010in. in thickness is classified as sheet. Sheet extrusion is generally used for stock up to 1/4 in. in thickness. The extruding system is the same used for other profiles, such as pipe and film, except for the die and the take-off equipment (Figure 38.3). Granular plastic is fed into the hopper of the extruder, and goes through the conventional extruding system. At the die, the soft, hot plastic is extruded directly into the finishing rolls at accurately controlled temperature. A conveyor system carries the sheet to the pull rolls, where it is fed to the cutting unit. The sheet is cut to specified lengths ready for packaging or carried directly to thermoforming units for further processing. Quality control and inspection are required to assure proper dimensional tolerances.

1. New words

pitch[pitʃ] *n.* 螺距,程度,斜度;*vt. & vi.* 定位于,用沥青涂,投掷,倾斜

lead[liːd] *n.* (螺杆)导程,导线,铅;*vt. & vi.* 引导,致使,导致

shank[ʃæŋk] *n.* 根部,后部,胫,腿骨

consistent[kən'sistənt] *adj.* 相等的

blending['blendiŋ] *n.* 共混

extrudate[iks'truːdeit] *n.* 挤出型材

coil[kɔil] *vt.* 卷取
thickness['θiknis] *n.* 厚度
lead[liːd] *n.* (螺杆)导程
adjustable[ə'dʒʌstəlb] *adj.* 可调节的

2. Phrases and expressions

feed section 加料段
compression section 压缩段
metering section 计量段
drive shank 驱动段
machined dies 机头,模头
profile shape 型材
irregular cross section 异型材
hard liner 硬衬套
heater bands 加热圈,电热圈(环)
wavy surface 波纹表面
non-uniform 非均匀性
take-off system 牵引系统
finishing roll 光辊
conveyor system 输送器,传送机系统
pull rolls 引出辊
dimensional tolerances 尺寸公差
extruder screw 挤出螺杆
upper die jaw 上模唇
restrictor bar 限流块
die land 口模区
chrome plated roll 镀铬辊
rubber covered 涂胶
pull-off roll 拖出辊
take-off unit 卷曲装置

3. Notes to the text

① Extruders are designated as to size by the diameter of the rotating screw. 挤出机是按照螺杆直径的尺寸进行标定的。

② The melted plastic finally passes through the screen pack, which removes dirt, and to the die at very high pressure. 熔融的塑料最后在非常高的压力下通过过滤污物的过滤网进入口模。

Lesson 39 Blow Molding

Blow molding is a process used to produce thin-wall hollow thermoplastic parts. A cylinder or tube of plastic, called a PARISON, is placed between the jaws of a mold. The mold is closed to pinch off the ends of the heated plastic and compressed air pressure is used to force the material against the mold faces①. When cool, the plastic becomes rigid(Figure 39. 1)

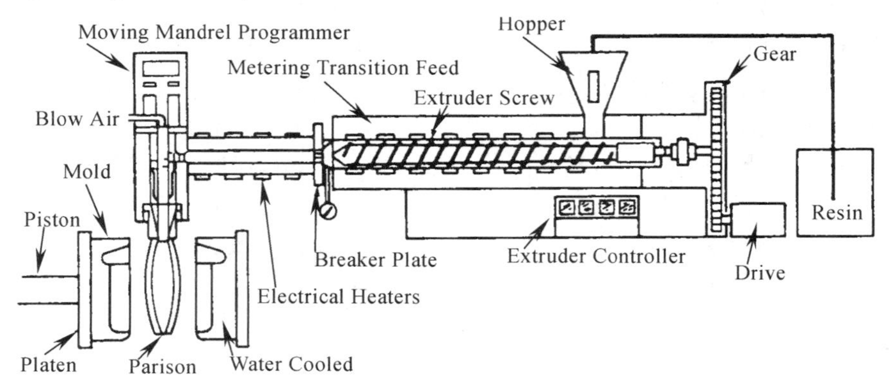

Figure 39. 1 Extrusion blow molding

The three main phases of the commercial process are:

(1) Softening the resin by the use of heat.

(2) Forming the parison.

(3) Blowing the parison in the mold.

The first of these phases involves the use of an extruder to heat the plastic to a molten state and compress the material at the die head. This part of the process is the same as typical extrusion. Second, the die or multiple dies, form the diameter and wall thickness of the extruded parison which is ready to be clamped between the mold halves. The third phase involves the closing of the mold halves by hydraulic pressure and pinching the parison. Air is used to expand the resin to conform to the mold cavity.

Figure 39. 2 This unusual twin bellows was blow molded as a single unit ready for use in a pumping system

In recent years blow molding has become one or the major processing methods of the plastics industry. With the rapid advancement of blow molding technology and machine design, the process is now to the stage of high-speed production with unique product design possibilities(Figure 39. 2). Almost any hollow object can be successfully blow molded, from chair seats and backs to automobile arm rests and sun visors.

Polyethylene resin is used extensively in blow molding. It is ideal for making a variety of products from soft, flexible squeeze bottles, to rigid containers.

Most thermoplastics can be blow molded. Ionomer, polyvinyl chloride, polycarbonate, and acetal resins are but a few that are used in considerable quantity. Perhaps the greatest single use of blow molding is in the production of disposable containers, such as milk bottles. They are lightweight, nonbreakable, and easily disposed of by incineration.

The size of a blow molding machine is determined by the extruder screw diameter, the number of die heads and the size of molds the machine will take. Blow molders are usually designed with the die heads lined up in a straight row with the molds on a fixed table, or a single die head on a multiple station mold turntable.

Of major importance in understanding the blow molding operation are the details of a typical mold. The parting line on the product appears where the mold halves meet. Those sections of the mold that squeeze the parison and weld it together prior to blowing are known as pinch-offs. The sections of the containers that have been pinched off are removed later during the trimming operation. The section pinched off at the bottom of a container is known as the tail. Machined aluminum is now the prime material for molds. Previously, extensive use was made of beryllium copper for the construction of the molds used in the blow molding process.

The sequence of the blow molding operation begins with the securing of the required molds onto the movable hydraulic platens (Figure 39. 3). The die is adjusted to extrude a parison of the estimated wall thickness, and the hopper is loaded with plastic granules either by hand or through an automatic vacuum system. The heaters are turned on for a warm-up period which softens the resin in the extruder cylinder and the die heads. Air pressure is adjusted for blowing the container and operating the strippers which eject the product. The water cooling system is turned on to maintain correct mold temperature and the hydraulic system is activated. This opens and closes the molds. The extruder is then turned on, set on automatic cycle, and a parison is extruded between the open halves of a mold. After the blowing cycle, the mold halves open and the container is ejected. In some operations, the container is conveyed to an automatic trimmer, while in others, the trimming operation is separate. As with most other thermoplastic processes, the scrap from trimming and defective containers can be reground and returned to the hopper to be used again. This sequence of the blow molding procedure, with variations, is characteristic of the procedure used in forming most blow molded products (Figure 39. 4).

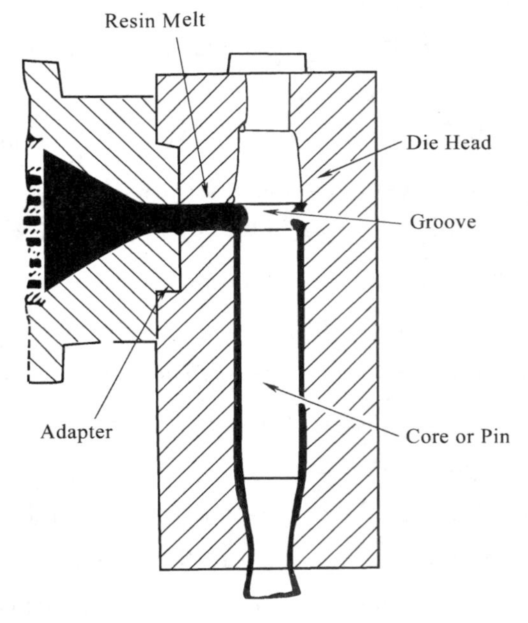

Figure 39. 3 Schematic drawing of side fed die with grooved care. The core is adjustable to give the required wall thickness to the parison

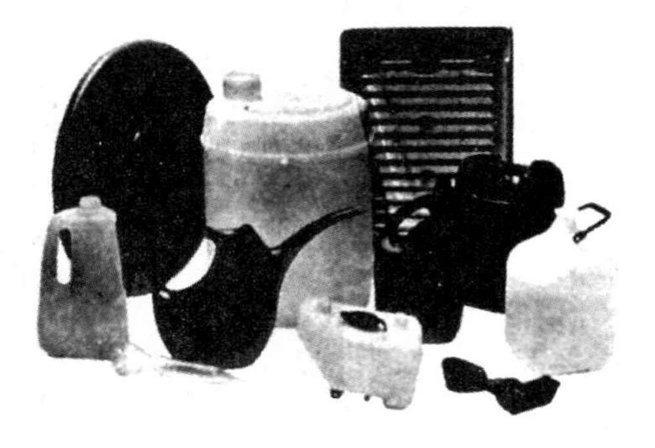

Figure 39. 4 An assortment of blow molded products

1. New words

hollow['hɔləu] *adj.* 中空的
parison['pærisən] *n.* 型坯
mandrel['mændrəl] *n.* 芯棒
jaw[dʒɔ:] *n.* 夹头
clamp[klæmp] *vt.* 合模
nonbreakable ['nɔnbreikəbl] *adj.* 不易碎的
disposable[dis'pəuzəbl] *adj.* 用完即可丢弃的
incineration[inˌsinə'reiʃən] *n.* 焚化,烧尽
turntable['tə:nteibl] *n.* 转盘
stripper['stripə] *n.* 脱模板
trimmer['trimə] *n.* 切边机
defective[di'fektiv] *adj.* 残次的
beryllium[bə'riljəm] *n.* 铍
groove[gru:v] *n.* 凹槽
adapter[ə'dæptə] *n.* 连接器

2. Phrases and expressions

thin-wall 薄壁的
pinch off 夹断,压紧
multiple dies 复合机头
sun visor 遮光板
mandrel programmer 芯棒控制装置
breaker plate 多孔板

3. Notes to the text

① The mold is closed to pinch off the ends of the heated plastic and compressed air pressure is used to force the material against the mold faces. 模具闭合将处于软化状态的型坯端切断,然后通入压缩空气使型坯完全贴紧模具的型腔。

4. Exercises

(1) A polymer used in blow molding is ________.

(2) The tube of hot plastic extruded between the molds of a blow molder is called a ________.

(3) The sections of a blown product where parts have been squeezed by the mold are known as ________.

(4) Three operations in the blow molding process are ________, ________ and ________.

(5) The size of a blow molding machine is determined by:

a. ________ b. ________ c. ________

(6) The metal most used for molds is ________.

a. aluminum b. steel c. brass

d. bronze e. none of the above

(7) What kind of objects may be successfully blow molded?

(8) What is the purpose of the stripper on a blow molding machine?

(9) Make a collection of hollow containers such as squeeze bottles and detergent bottles. Examine each container to determine if it was blow molded by checking these factors:

a. Look for a weld line at the bottom of the container where the tail may have been pinched off.

b. Check the neck area of the container where trimming may have been necessary.

c. See if the neck has been machined to seat the cap.

d. Check for a parting line around the product where the mold halves came together.

Reading Material

Blown Film, Flat and Wire Coating Extrusion

Blown Film: Blown film extrusion is similar to the other extrusion processes except the die forms a hollow tube through which air is forced to expand the film into a cylinder, called a "bubble". As the plastic bubble solidifies, it is squeezed together between rolls to form a double thickness film (Figure 39. 5). It has been found satisfactory to extrude and pull the bubble upward, flattening it between rolls, and carrying it to a wind-up roll. In some cases the film is cut to short lengths and sealed at one end to form plastic bags. It can also be split lengthwise and used for larger sheeting.

Wire coating: The plastic coating of wire is another major use of the extrusion process. It is similar to the extrusion of pipe or tubing except the mandrel in the die is replaced by a tapered guide through which a continuous line of wire is fed (Figure 39.6). As the plastic flows from the extruder through the die, it surrounds the moving wire which is preheated to the plastic melt temperature and leaves the die as a integral unit①.

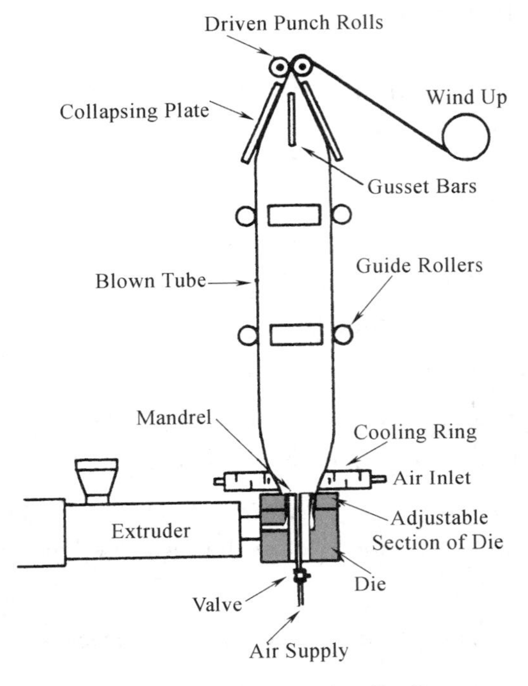

Figure 39.5 Blow extrusion film line

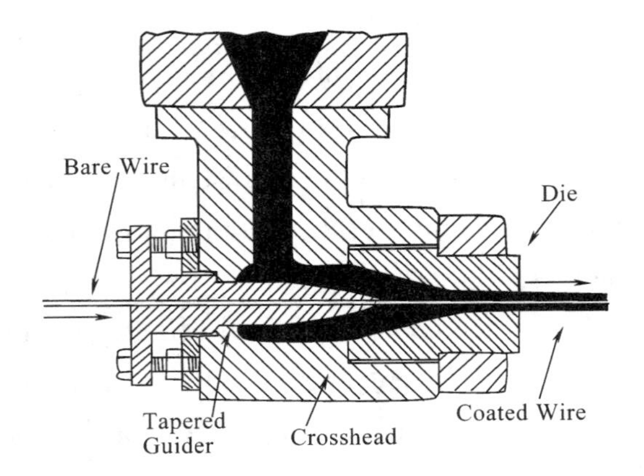

Figure 39.6 A crosshead holds the wire-coating die and the tapered guider as the soft plastic flows around the moving wire

Many thermoplastic resins are used in wire and cable coating. The polyethylenes, polyvinyl chloride, and nylon are typical wire coating resins. Silicon is often used for high heat resistant applications. Figure 39.7 shows the complete line in a wire coating system.

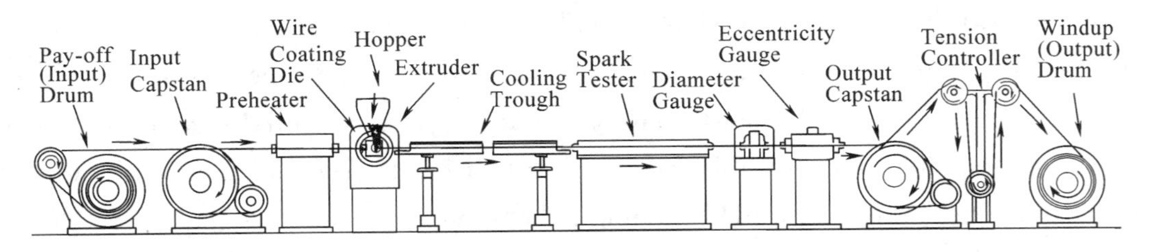

Figure 39.7 A general layout of the components in a wire-coating extrusion plant

Compounding and Granulating: In a resin compounding plant, the extruder is used for the blending, coloring, and granulating of resin to be shipped to the processor. The plastic resins often require additives for particular applications along with the required colors desired by the processor. These are all added to the resin and fed into the extruder to provide a homogeneous mix in the resulting granules. A special granulate die is used on the extruder (Figure 39.8), which produces many strands of resin. These strands are then cut into small pieces approximately 1/8 in. in length and pulled into loading tanks through an exhaust system. The granulated resin is normally packaged in fifty pound bags or one thousand pound cartons ready for shipment. Typical granulated resin is in the form of round, cylindrical, or cube shaped particles (Figure 39.9).

Figure 39.8 This hot melt granulator shows the many strands of plastic being extruded through the die into the grinder below

Figure 39.9 Plastic granules from extrusion process ready for molding

Resin coloring is done either by adding dry powder color during the compounding process or by the use of color concentrates at the time of molding. Color concentrates appear as small granules similar to the resin except that they are almost pure color pigment bonded together by a small amount of the same resin used in molding[②]. The normal coloring ratio is 5 lbs. of color concentrate to one-hundred pounds of natural resin.

Regrinding of thermoplastic materials from the extrusion process, as well as other molding processes, makes use of scrap and reject resin. Trimmings from extrusions may be reground and remolded. The reground material is usually mixed with new material of the same grade and color.

1. New words

bubble['bʌbl] *n.* 膜泡

solidify[sə'lidifai] *vt.* 使凝固，硬化

split[split] *n.* 裂口，薄片

lengthwise['leŋθwaiz] *adj.* & *adv.* 纵向的(地)

mandrel['mændrəl] *n.* 芯模

crosshead['krɔshed] *n.* 十字头，丁字头，小标题，子题

granulating['gænju'leitiŋ] *n.* 造粒

strand[strænd] *n.* 线料

concentrate['kɔnsentreit]n. 浓缩物，母料
pigment['pigmənt]n. 颜料
regrinding[ri'graindiŋ]v. & n. 再粉碎
regrind[ri'graund]v. 再粉碎，再研磨
gusset['gʌsit]n. 边折，衣袖
collapse[kə'læps]v. 压扁[平]，毁坏，断裂
bare[bɛə]adj. & vt. 赤裸的，无遮蔽的，空的，使赤裸，露出
taper['teipə]n. 锥形，锥度；v. 逐渐变细，逐渐减少
capstan['kæpstən]n. 绞盘，起锚机
trough['trɔ:f]n. 槽，水槽，饲料槽，木钵
spark[spa:k]n. & vi. & vt. 火花
gauge[gedʒ]n. 标准尺，规格，量规，量表； v. 测量
eccentricity[eksen'trisiti]n. 偏心，古怪，[数]离心率
windup['waindʌp]n. 收卷，卷起

2. Phrases and expressions

tapered guide　锥形导线器
integral unit　整体部件
driven punch roll　传动轧辊
collapsing plate　压扁[平]板，夹膜板，人字板
blown tube　吹胀膜管，管膜
gusset bar　(吹膜机)边折板
guide roller　导向辊
cable coating　电缆涂层
bare wire　裸线
pay-off drum　送料卷盘
input capstan　送料(输入)绞盘
cooling trough　冷却槽
spark tester　电火花实验(检测)
diameter gauge　直径测量
eccentricity gauge　偏心测量
tension controller　张力控制器
windup drum　卷取卷盘
loading tank　加料槽
exhaust system　排气系统
cube shaped　方形的
color concentrate　色母料(粒)

3. Notes to the text

① As the plastic flows from the extruder through the die, it surrounds the moving wire which is preheated to the plastic melt temperature and leaves the die as a integral unit. 当塑料熔体流过机头时，塑料熔体就自然地包覆在已预热成塑料熔体温度的移动电线周围，并作为一个整体离开口模。

② Color concentrates appear as small granules similar to the resin except that they are almost pure color pigment bonded together by a small amount of the same resin used in molding. 色母料看上去是和树脂差不多的小颗粒，但实际上它们是由少量的模塑所用的同类树脂粘连在一起的近乎纯粹的颜料。

Lesson 40 Calendering

Calendering is the process of squeezing a softened thermoplastic material between two or more rolls to form a continuous film. The process, originally adapted from the rubber industry, is a major method of producing plastic film and sheet.

The material generally used in the calendering process is flexible polyvinyl chloride. However, some film is calendered from ABS, cellulosics, polyethylene, and polystyrene.

Figure 40. 1 shows the stages in the calendaring process. The thermoplastic resin may be mixed with lubricants, stabilizers, plasticizers, and colorants. During mixing, the mass is heated and becomes rubbery, like hot, soft clay. It is then fed into heated calendering rolls, and squeezed to the desired thickness as it passes through the rolls. Passing from the calendering rolls, the vinyl film goes through cooling rolls, is cut to width by the edge trimmer, and is wound on a take-off roll. Figure 40. 2 shows the resin flow as it passes through the calendering rolls. Thickness of the calendered sheet is closely controlled by the space between the final rolls.

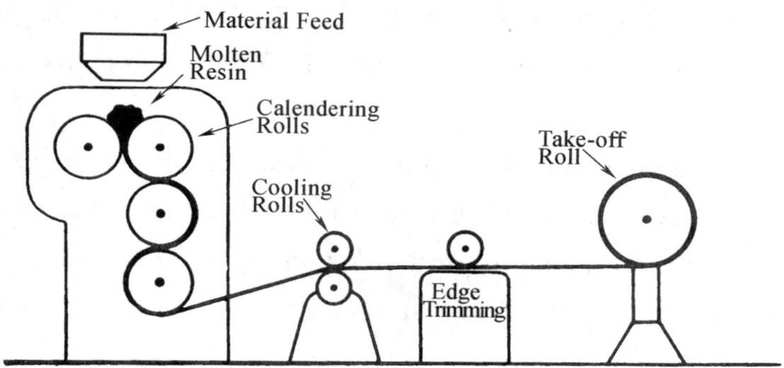

Figure 40. 1 Schematic illustrating the main stages in the calendering of polyvinyl chloride film

The equipment used in calendering consists of four main units. A mixer or mill is used to blend the ingredients into the proper compound for processing. The major unit is the calendering machine (Figure 40. 3), which compresses and rolls the plastic into a flat sheet or film. Finishing equipment includes cooling rolls and an edge trimmer. Finally, after finishing, the sheet is wound onto a take-off roll.

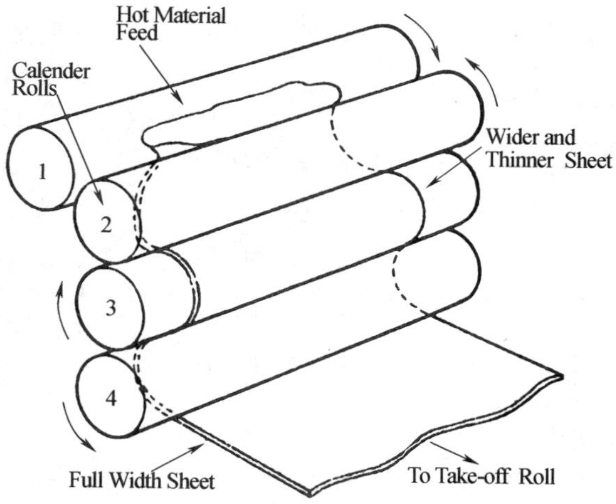

Figure 40. 2 Widening and thinning of calendered sheet as it passes through the rolls

Figure 40. 3 This huge calender has four rolls 66 in. wide. It is called a four-roll inverted L because of the location of the rolls as seen in Figure 40. 1

Figure 40. 4 Small two-roll calender having a maximum width of 16 in

The calender itself serves to slowly reduce the plastic thickness as each set of rolls is spaced closer together①. The rolls are heated to maintain forming temperature and are powered to draw the plastic through. Patterned or textured sheet is produced by embossed rolls. A small calendering machine, suitable for laboratory use, is shown in Figure 40. 4.

Calenders are also used to coat sheet materials such as paper and fabric. The process is similar to regular calendering except fabric is fed into the calendering rolls as the plastic sheet is being formed. Hot and soft plastic is forced against the fabric. The plastic and fabric are tightly bonded together, emerging as a single sheet through the final rolls.

Calendering is used to produce products such as automotive and furniture upholstery, shower curtains, clear film for packaging, rain coats and floor coverings.

1. New words

lubricant['lju:brikənt] *n.* 润滑剂
clay[klei] *n.* 黏土
mixer['miksə] *n.* 混炼机，混合机
mill[mil] *n.* 研磨机
draw[drɔ] *v.* 拉，拽
fabric['fæbrik] *n.* 织物，纤维织物
emboss[im'bɔs] *vt.* 压纹，压花

2. Phrases and expressions

calendering roll 压延辊
cooling roll 冷却辊
finishing equipment 压光机
embossed roll 压花辊
take-off roll 牵引辊
patten roll 压花辊(模型辊)
hot material feed 热材料供料
wider and thinner sheet 片材变宽变薄

3. Notes to the text

① The calender itself serves to slowly reduce the plastic thickness as each set of rolls is spaced closer together. 当每组压延辊间的间隙逐渐减小时，压延机本身即起到了逐渐减小塑料片材厚度的作用。

4. Exercises

(1) Calendering is used to produce plastic ________ and ________.

(2) Common plastic resins used in the calendering process are

a. ______. b. ______. c. ______. d. ______.

e. ______.

(3) Patterned material is produced by using ________ rolls.

(4) The calendering process was originally adapted from the ________ industry.

(5) Materials such as _______ and _______ are coated by the calendering process.

(6) The four main units in calendering equipment are

a. ______. b. ______. c. ______. d. ______.

(7) Five common products made by the calendering process are

a. _____. b. _____. c. _____. d. _____. e. _____.

Reading Material

Laminating Molding

A laminated material is made by bonding together two or more layers of material to form a single unit or sheet. Plywood is a good example of such a material. Sheets of wood veneer are glued together to form a thicker, more rigid sheet. The same principle is involved in laminating plastics.

In laminating plastics, the individual layers may be sheet plastic; or layers of other materials such as paper, fabric, and wood may be included with the plastic.

The individual layers of the laminate may be bonded by using a synthetic adhesive, or by fusion of the layers of plastic. In another type of laminate, individual layers are impregnated with a synthetic resin which bonds the layers together.

A laminate formed by pressures exceeding 1,000 psi is known as a high-pressure laminate. Formica used for kitchen cabinet tops is a good example of a high-pressure laminate. Laminates involving molding pressures of 0 to 1,000 psi are known as low-pressure laminates. Laminated credit and identification cards are examples of low-pressure laminates.

Most laminating is done in a heated platen hydraulic press(Figure 40.5). The layers of material are usually impregnated with a thermosetting resin and allowed to dry. Typical resins used for impregnating are the phenolics, melamines, silicones, epoxies, and polyesters. The impregnated layers are assembled to the proper thickness and place in the laminating press between polished plates(Figure 40.6). As heat and pressure are applied, the resin flows throughout the layers of material forming a solid laminate mass①. Laminating temperatures range from 300 to 350 ℉. Pressures range from 1,000 to 2,000 psi. When thermosetting resins are used, the laminate cures in the press in a few minutes. It can be removed from the press while still hot.

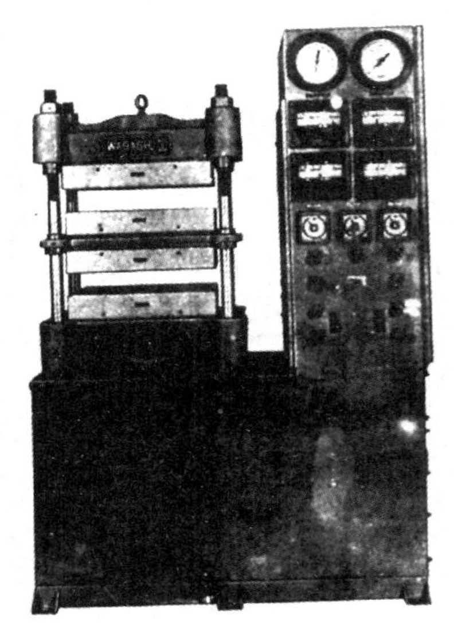

Figure 40.5 Double heated platen hydraulic laminating press

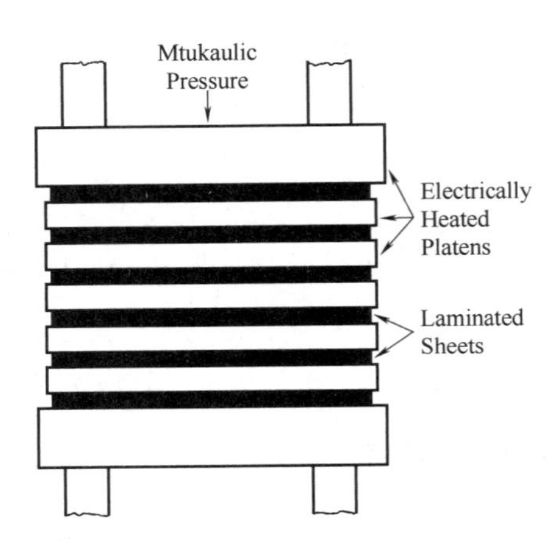

Figure 40.6 Multiple high-pressure laminating in a hydraulic press

Laminating thermoplastic sheet is done in a similar manner except that after the sheets are fused together by using heat and pressure, the platens must be cooled before removing the product. Longer cycle periods are required as the press must be reheated before the next laminate can be prepared.

Materials most commonly laminated are flat sheets. These sheets are usually prepared for specific purposes, such as counter tops, in which a number of layers of paper with a colored top sheet are impregnated with phenolic or melamine resin and laminated into a hard, durable material. Paper is used extensively because of its toughness and relatively low cost. Cloth fabrics, such as canvas, are often used for sheet laminates with phenolics and epoxies for electrical insulating pieces, transformer insulation, fuse boxes, and terminal blocks. Such laminated sheet is also machined or punched to form cams and silent running gears.

Decorative laminates are used for furniture, wall and ceiling panels, and display cases. Laminated thermoplastic sheets are used to protect photographs and identification papers, notebook covers, and credit cards. Some rod and tubing stock is laminated to provide extra strength and electrical resistance.

1. New words

formica[fɔːmaikə] *n.* 胶木

canvas['kænvəs] *n.* 帆布

2. Phrases and expressions

identification card 身份证

flat sheet （平）片材

cloth fabric （布）织物

hydraulic pressure 液压

3. Notes to the text

① As heat and pressure are applied, the resin flows through the layers of material forming a solid laminate mass. 树脂受热和受压后即开始在一层层材料内流过，形成一个固体的层压板材。

Lesson 41 Plastic Foam Molding

Foamed plastics (also called expanded or cellular plastics) are made by adding air or gas to a plastic resin to form a sponge-like material. Modern developments in plastic foam molding have placed this among the major processing techniques of the industry. Resins commonly used in the production of plastic foam include polystyrene, polyurethane, polyethylene, cellulose acetate, epoxy, silicone, and phenolic.

Plastic foams may be classified as to material, type, cell structure and density. The two principal types of plastic foam are rigid and flexible. Rigid foams are resistant to crushing, while flexible foams are easily crushed.

Cell structure refers to the openings in the foam. Plastic foams may have either an open or closed cell structure. Open cell structure indicates the foam has interconnected cells or openings running through the material. Foam with an open cell structure will allow a liquid to pass through like a sponge. This kind of foam structure is usually flexible. Closed cellular structure means the cells within the plastic foam are separate air or gas filled units. The closed structure will not allow liquid to pass through the foam. Closed structure foams are usually rigid.

The density of plastic foam is indicated by its weight in pounds per cubic foot. Density of foams ranges from 0. 10 to 70. 0 lbs. /cu. ft. The density of a foam is important for product uses in terms of weight, insulating properties, and flotation. A number of plastic foams are shown in Figure 41. 1.

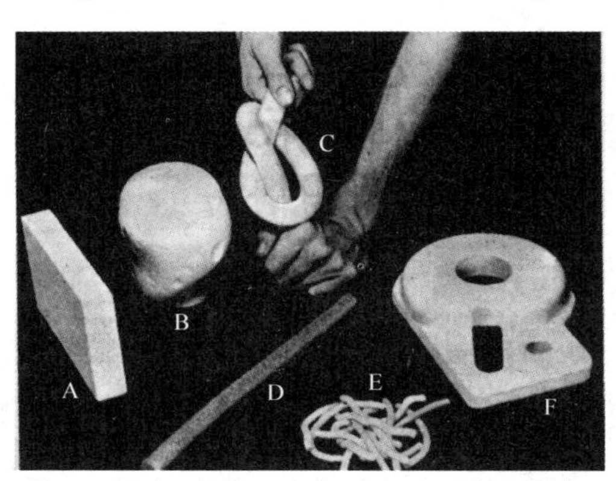

Figure 41. 1 A-Extruded polystyrene foam slab. B-Rigid polyurethane foam. C-Flexible polyethylene foam rod. D-rigid cellulose acetate foamed rod. E-Expanded polystyrene foamed spaghetti. F-Molded polystyrene foam

Plastic foams are produced by three principal methods: mechanical, physical, and chemical

In MECHANICAL FOAMING of a plastic resin the action is similar to preparing a milk shake. The resin, as a solution or emulsion in liquid form, is vigorously agitated until it becomes foam of air bubbles. Fusing of the foamed resin by heat causes it to remain solid foam. Polyvinyl chloride plastisols can be mechanically foamed in this manner.

PHYSICAL FOAMING usually involves the use of compressed gasses or chemicals which change their physical form during the foaming process. Foamed polyethylene, for example, is produced by forcing compressed nitrogen gas into molten polyethylene while the resin is under pressure. As the pressure on the molten resin is released, either in a mold or an extruder, the nitrogen expands and foams the plastic①.

CHEMICAL FOAMING, in simple terms, involves mixing or dissolving a chemical compound in a liquid resin which will react, usually under heat, to form a gas. The gas released within the molten or liquid resin causes the foaming action②. Foamed polyurethanes are produced by this method.

Polystyrene is the principal resin involved in the manufacture of expandable beads. Gas is contained in each tiny, hollow pellet or bead. When heat is applied, the gas or blowing agent causes the bead to expand. At the same time, the styrene skin softens, allowing the bead to blow up like a small balloon. Since the skin is soft and molten as the bead expands, it will fuse to other beads. If enough beads are placed in a confined mold and heated, they will expand to fill the mold cavity and provide enough pressure to "weld" the bead together to form a rigid foam. This is closed cell structure and, after molding, is called an expanded foam.

Expandable polystyrene beads are available in various graded sizes. The larger sizes are used for molded blocks and packaging cases, and the smaller beads for articles like drinking cups. Molds are usually made of aluminum or stainless steel. On a production basis, they are mounted in presses which open and close for loading and fusion cycles.

A typical molding sequence includes pre-expansion of the beads, filling the mold with pre-expanded beads, subjecting the mold to heat, cooling, mold opening, and ejection of the product.

1. New words

cell[sel] *n.* 泡孔
crush[krʌʃ] *vt.* 压碎,压裂
indicate['indikeit] *vt.* 指示 表明
sponge[spʌndʒ] *n.* 海绵
shake[ʃeik] *n.* 摇动,震动
vigorously['vigərəsli] *adv.* 充满活力地
agitate['ædʒiteit] *vt.* 搅拌
fusing['fjuːziŋ] *n.* 熔融,塑化
expand[iks'pænd] *n.* 膨胀
frame[freim] n. 模架
expansion[iks'pænʃən] *n.* 膨胀
bead[biːd] *n.* 珠粒

2. Phrases and expressions

formed plastics 泡沫塑料
cellular plastics 泡沫塑料
sponge-like material 类海绵材料
cell structure 泡孔结构
resistance to crushing 抗压裂性
closed cell structure 闭孔结构
open cell structure 开孔结构
interconnected cell 联泡孔
cellular structure 微孔结构
physical foaming 物理发泡
foaming process 发泡过程
foaming action 发泡作用

expandable bead　可发珠粒料
blowing agent　发泡剂
hollow pellet　空心粒料
molded block　模体,模块
stainless steel　不锈钢
fusion cycle　塑化周期

3. Notes to the text

① As the pressure on the molten resin is released, either in a mold or an extruder, the nitrogen expands and foams the plastic.　当压力在模具中或在挤出机中的熔融树脂中卸压时,氮气膨胀使塑料发泡。

② Chemical foaming, in simple terms, involves mixing or dissolving a chemical compound in a liquid resin which will react, usually under heat, to form a gas. The gas released within the molten or liquid resin causes the foaming action.　简单地说,化学发泡通常是在热的作用下在液体树脂中混合或溶解一种能够与树脂相互反应而形成气体的化合物。在熔融体或液体树脂中释放的气体引起发泡。

4. Exercises

(1) Plastic foams have either an ________ or a ________ cell structure.

(2) Foamed plastics are also called ________ or ________ plastics.

(3) Plastic foams are classified as to:

a. ________　b. ________　c. ________　d. ________.

(4) The type of foam refers to its properties of being either ________ or ________.

(5) Which of these plastics are used for foaming?

a. Polyurethane.　b. Polyethylene.

c. Cellulose acetate.　d. All of the above.

e. None of the above.

(6) The density of plastic foam is indicated by ________.

(7) The three methods of producing foams are ________ and ________.

Reading Material

Casting Molding

Casting refers to any of a number of processes in which a liquid plastic is poured into a suitable mold. The plastic may be in the form of a monomer which is polymerized after it is poured, or a liquid resin to which a catalyst is added prior to pouring. The molds range from complicated to simple home workshop types. See Figure 41. 2. Resins can be cast cold or hot and allowed to solidify through further polymerization with heat added if necessary.

Common casting resins are the acrylics, polyesters, silicones, epoxies, phenolics, and polyurethanes. Resins are often filled for reinforcement. They are available from clear to opaque and may be colored to any desired shade.

Molds used in casting processes are usually inexpensive and are made from materials such as plaster, glass, wood, metal, and other plastics. A major use of casting in the plastics industry is to produce cast acrylic sheet, tubing, and rods. Phenolic billiard balls, jewelry, and imitation marble are made by casting. The polyesters are used for embedding scientific specimens, hobby work, cast sheets, furniture parts and jewelry. Epoxy casting resins are used extensively in the automotive and aircraft industries for making molds, tooling jigs, and dies. Polyurethane is used in both flexible and rigid casting for items such as furniture parts, automotive crash padding, and housewares. The flexibility of silicone resins has made them ideal for casting molds into which other resins are cast.

Figure 41.2 Crystal clear acrylic are made by casting an acrylic monomer syrup in glass mold. Here they are being placed in a heating chamber for curing

One of the major casting processes, that of producing cast acrylic sheet, is done by using a premixed acrylic monomer in the form of a thick syrup. The liquid is poured between two sheets of polished glass and allowed to cure. After casting, the sheets are heated to relieve any stresses, cut to size, and paper coated to protect the polished surfaces.

Another production casting process makes use of polyester casting resins (Figure 41.3). The resin and catalyst are mixed automatically in a dispenser unit and the liquid polyester is ready for continuous casting. Production furniture and cabinet parts, made by casting in this manner, closely resemble wood.

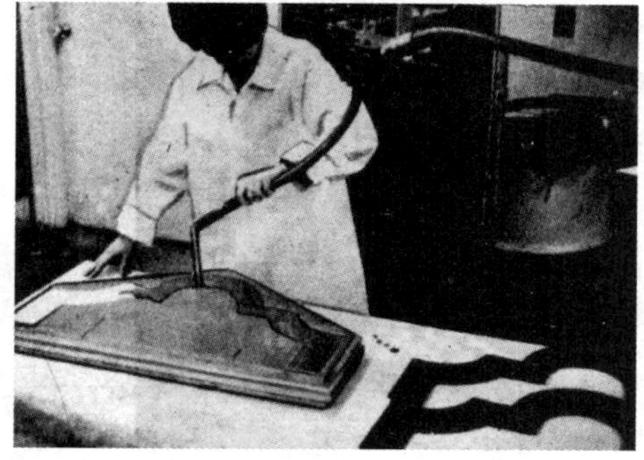

Figure 41.3 Polyester casting resin is poured into a mode in the production of furniture parts as shown on the right

Silicone resins are used extensively for making molds into which other plastics are cast. For example, silicone resin is poured into a mold in which a pattern of a picture frame is fastened. The resin is poured carefully over parts of the pattern to avoid trapping air bubbles. When the mold is completely filled with resin, the cover, which has been coated with a mold release, is placed over the mold and fastened to the mold box①. After a 24 hour curing period, the cover is removed and the silicone mold is stripped from the pattern and placed in a mold box ready to serve as a mold to be used in casting a rigid polyurethane picture frame with considerable detail②. The rigid polyurethane resin is then poured into the silicone mold, allowed to cure, and removed from the flexible mold. The edges of the frame are trimmed and sprayed with a finish. The silicone mold has an average life of 100 to 200 products.

Because of its high heat resistance and ability to reproduce exacting detail, cast silicone molds are often used to make metal prototype parts. The mold is prepared by casting the silicone resin around a clay pattern. After curing, the pattern is removed from the silicone mold and a lead alloy is poured into the mold. The lead casting is easily removed and another can be poured right away. Hundreds of prototype parts can be produced with little mold deterioration.

1. New words

plaster['pla:stə] *n.* 石膏； *vt.* 涂以灰泥，敷以膏药
marble['ma:bl] *n.* 大理石
hobby['hɔbi] *n.* 切压
jewelry['dʒuəlri] *n.* 宝石，珍宝
jig[dʒig] *n.* 夹具
syrup['sirəp] *n.* 树脂浆
dispenser[dis'pensə] *n.* 喷洒器
prototype['prəutətaip] *n.* 原型
curing['kjuəriŋ] *n.* 固化，熟化，硫化

2. Phrases and expressions

casting molding 铸塑模塑，铸塑成型
acrylic sheet 丙烯酸酯类片材
imitation marble 仿大理石
prototype part 原型零件
picture frame 画框

3. Notes to the text

① When the mold is completely filled with resin, the cover, which has been coated with a mold release, is placed over the mold and fastened to the mold box. 当树脂充满模具时，将一个带有排气孔并涂上树脂的盖盖住模具并将其锁紧到模盒上。

② After a 24 hour curing period, the cover is removed and the silicone mold is stripped from the pattern and placed in a mold box ready to serve as a mold to be used in casting a rigid polyurethane picture frame with considerable detail. 24h 后硅树脂固化，然后将上盖移开，再将硅模从图案模上取下，放到一个模盒中，此硅模即可用来制作铸塑硬聚氨酯的精美画框。

Lesson 42 Thermoforming

Thermoforming plastic sheet material in sheet thermoforming is one of the major processing techniques of the plastics industry and also one of the oldest. Since early attempts, about the turn of the century, to shape cellulosic sheet, the process has grown rapidly. Much of the growth is due to innovative techniques in forming equipment and to the development of new sheet materials with special thermoforming properties.

Thermoforming is the heating of sheet plastic to a pliable state and forcing it around the contours of a mold by using pressure. The required pressure is usually mechanical or air pressure, often assisted by localized heating and bending. There are three main classifications of thermoforming with numerous variations within each category: Matched Mold Forming, Vacuum Forming, and Pressure Forming.

MATCHED MOLD FORMING

One of the more exacting techniques of thermoforming plastic sheet material, matched mold forming, requires two mold halves that fit together perfectly. The thermoplastic material is heated to its softening point and formed by mechanical pressure between the two halves of the mold. Since both halves of the mold contact each surface of the softened sheet, it is necessary that the mold surfaces be highly polished or textured to product requirements.

The molds are usually made of aluminum or steel and mounted in a hydraulic or pneumatic press. Heated sheet material is placed between the molds and the press is closed. The mold is normally water-cooled to control mold temperatures. Matched mold forming is especially suitable to products requiring excellent reproduction detail.

VACUUM FORMING

Of all of the thermoforming processes, vacuum forming is the most versatile. It consists of heating a plastic sheet, held in a frame until it is soft and pliable. The mold is placed directly under the sheet and slight pressure is applied to seal the plastic to the upper mold edge[①]. A vacuum is applied through small holes in the mold cavity and the atmospheric pressure forces the softened sheet against the contour of the mold walls, Figure 42. 1. Upon cooling, the product has solidified and when removed it retains the shape of the mold.

Vacuum forming is generally the least expensive of the thermoforming processes because the mold is made of one piece and is of relatively simple construction. Molds can be made of cast aluminum, machined aluminum, cast filled epoxy, wood, or plas-

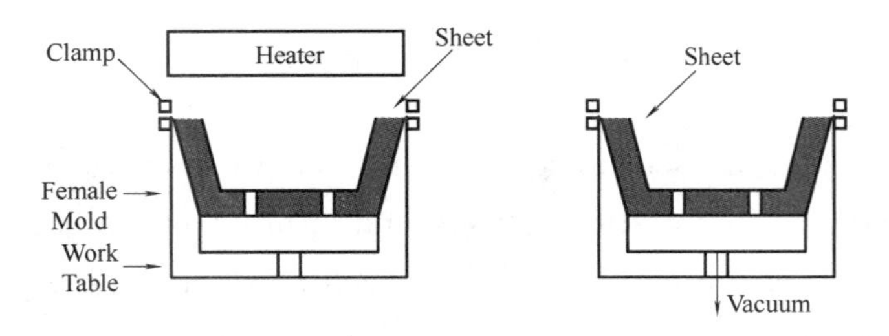

Figure 42. 1 Straight vacuum forming in a female mold is recommended for low-profile parts where deep draw is not a requirement

ter. Mold material selection is usually based on the production run. Hardwoods and plaster are ideal for prototype work and short production runs. Metal molds will produce large quantities of products with but little wear.

FORMING PROCESS

Many variations of straight vacuum forming have been developed to provide for a more even stretch of the material and a consistency of wall thickness on deep-drawn products[2]. These include snap-back forming, Figure 42. 2, where a vacuum is drawn on the softened sheet to evenly stretch it down into the vacuum box. At the same time, a male mold is lowered into the cavity, the vacuum released, and the sheet quickly conforms to the contour of the male mold.

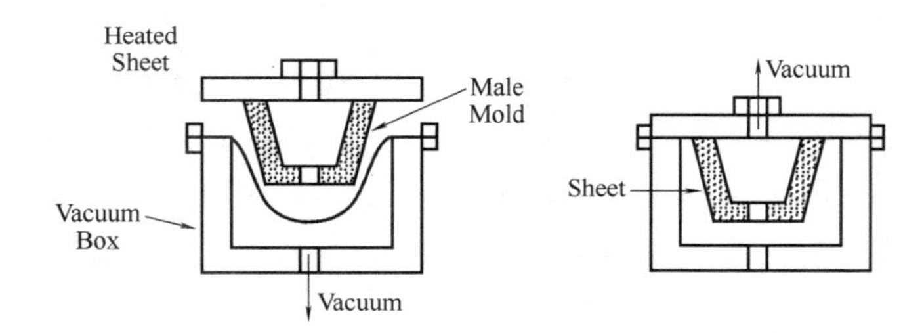

Figure 42. 2 Vacuum forming with snap-back can reduce starting sheet size, aid material distribution, and minimize chill marks

Following the sequence through the vacuum forming process, the selected sheet material is cut to size and placed in the clamping frame. The sheet is heated to the correct forming temperature and the mold is brought into position for the vacuum draw as shown in Figure 42. 2. A short cooling period allows the sheet to solidify and the product is removed from the mold. The product is then trimmed from the surrounding sheet by stamping, slitting, or sawing.

The versatility of vacuum forming makes it suitable for extremely large products, as shown in Figure 42. 3, and high production of smaller items.

PRESSURE FORMING

The third main process in thermoforming is known as pressure forming; sometimes called air blowing. The process should not be confused with blow molding since it deals only with the forming of sheet plastic. The two primary techniques are Straight Pressure Forming and Free Blowing.

Figure 42.3 Sheets of ABS-polycarbonate alloy 1/4 inch thick are used to vacuum form this complete automobile body. A body can be formed in a 20 minute cycle

STRAIGHT PRESSURE FORMING involves a female mold over which a sheet of thermoplastic material is clamped, as shown in Figure 42.4. A radiant heater softens the sheet, a cover is quickly placed over the hot sheet, and preheated compressed air is blown through the cover opening. The sheet is forced against the contours of the mold and any air trapped below the sheet escapes through vent holes in the mold③. After cooling, the formed part is removed from the mold and is trimmed similar to vacuum forming.

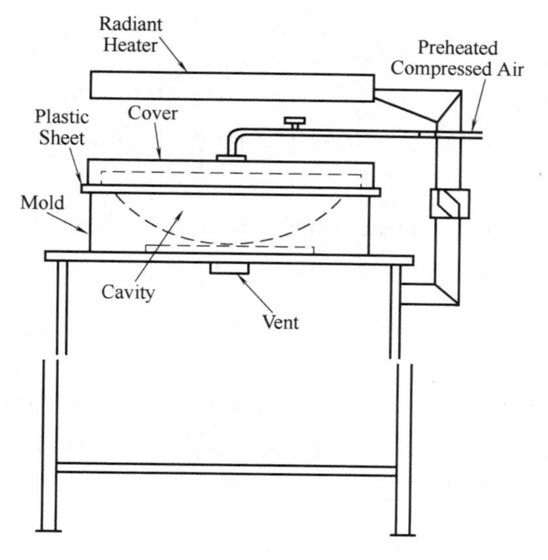

Figure 42.4 Details of a sheet pressure forming machine

1. New words

cellulosic[ˌselju'ləusik] *adj.* 纤维素的，有纤维质的

innovative['inəuveitiv] *adj.* 新发明的，新引进的，革新的

pliable['plaiəbl] *adj.* 易弯的，柔韧的，柔顺的

contour['kɔntuə] *n.* 外形，轮廓

bending['bendiŋ] *n.* 弯曲（度），挠度

exacting[ig'zæktiŋ] *adj.* 需细致小心的,(标准)严格的,难达到的

polished['pɔliʃt] *adj.* 擦亮的,抛光的

aluminum[ˌælju'miniəm] *n.* 铝

pneumatic[nju(:)'mætik] *n.* &*adj.* 气胎;气动的

water-cooled['wɔ:təku:ld] *adj.* 用水冷却的

versatile['və:sətail] *adj.* (指工具、机器等)多用途的,多功能的

cavity['kæviti] *n.* 洞,空穴,腔

deep-drawn[ˌdi:p'drɔ:n] *adj.* 深长的

stamping['stæmpiŋ] *v.* 冲压; *n.* 压膜

slit[slit] *vt.* 切开

sawing['sɔ:iŋ] *n.* 锯,锯切,锯开

radiant['reidiənt] *adj.* 放热的,发光的,辐射的

2. Phrases and expressions

pressure forming 压力成型法

matched-mold forming 对模成型

mold halves 模具组分

softening point 软化点,软化温度

pneumatic press 气动压力机,气压机

hydraulic press 液压机

cast aluminum 生铝,铸铝

snap-back forming (热成型)快速吸塑成型,骤缩成型

male mold 阳模

straight pressure forming 直接压力成型

free blowing (热成型)自由吹胀成型(即无模吹气成型)

female mold 阴模,下半模,凹模

vent hole 排气孔,通风孔

3. Notes to the text

① The mold is placed directly under the sheet and slight pressure is applied to seal the plastic to the upper mold edge. 模具放在薄片的正下方并施加微小的压力将塑料封在模具的上表面上。

② Many variations of straight vacuum forming have been developed to provide for a more even stretch of the material and a consistency of wall thickness on deep-drawn products. 从直接真空成型法又演变出很多不同的方法,目的是为了获得更加均匀的材料延展性以及深长形制品壁厚的均匀性。

③ The sheet is forced against the contours of the mold and any air trapped below the sheet escapes through vent holes in the mold. 薄片材被紧贴在模具的轮廓上,所有被困在薄片下面的气体都从模具的排气孔中逸出。

4. Exercises

(1) What are the features of the three main categories of thermoforming? What are their separate applications?

(2) How do you give a plastic sheet the required shape in vacuum forming?

(3) What is the sequence through the vacuum forming process?

(4) What is pressure forming and what makes it different from blow molding?

(5) Translate the following sentence into Chinese:

Since both halves of the mold contact each surface of the softened sheet, it is necessary that the mold surfaces be highly polished or textured to product requirements.

Reading Material

Material Compounding of Polymer

Polymers are almost always used in combination with other ingredients. These ingredients are discussed in subsequent chapters, but they must be combined with the polymer in a compounding operation.

Occasionally, if they don't interfere with the polymerization reaction, such ingredients may be incorporated at the monomer or low molecular weight polymer stage and carried through the polymerization and/or crosslinking reaction. In such cases, viscosities are low enough to permit the use of standard mixing equipment. Similarly, powdered PVC and thermosets are compounded with other ingredients in the usual tumbling-type of blending equipment.

Because of their extremely high melt viscosities, specialized equipment is usually needed to compound ingredients with high molecular weight thermoplastics, however①. In general, high shear rates and large power inputs per unit volume of material are required to achieve a uniform and intimate dispersion of ingredients in the melt.

Single-and twin-screw extruders are used extensively for continuous compounding. The latter, with screws often modified to incorporate special mixing sections, are better mixers and provide a narrower, more uniform distribution of residence times, while the former offer lower cost and greater mechanical simplicity. Intensive mixers, such as the Banbury (Figure 42. 5), subject the material to high shear rates and large power inputs in a closed, heated chamber containing rotating, intermeshing blades.

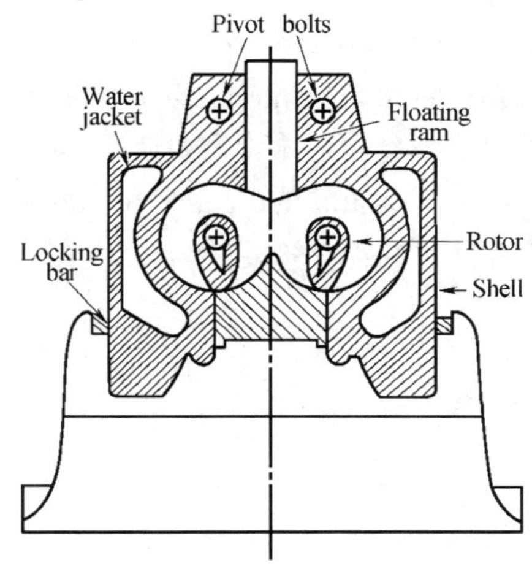

Figure 42. 5 Banbury mixer

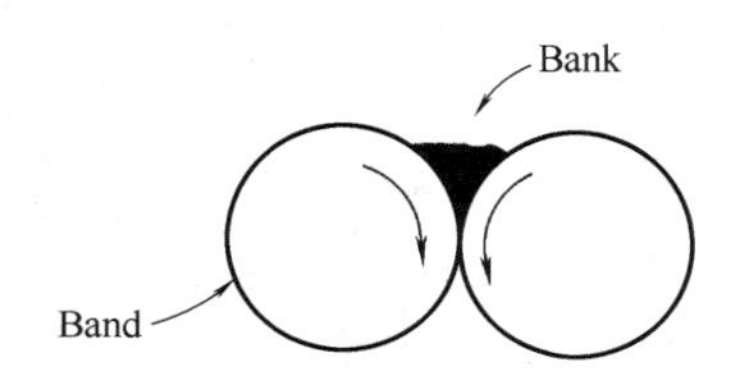

Figure 42. 6 Two-roll mill

A two-roll mill(Figure 42.6) generates high shear rates in a narrow nip between two heated rolls which counter-rotate with slightly different velocities. In commercial mills, the rolls are about 1 ft in diameter by 3 ft long. Once the polymer has banded on one of the rolls, ingredients are added to the bank between the rolls. The band is cut off the roll with a knife, rolled up, and fed back to the nip at right angles to its former direction②. This is done several times to improve mixing. Despite extensive safety precautions, operators of two-roll mills often have n fingers($n<10$). To the author's knowledge, no systematic studies of the effect of the additional($10-n$) ingredients on the properties of the compounded polymers have been reported.

1. New words

ingredient[in'gri:djənt]n.(混合物的)组成部分,配料

incorporated[in'kɔ:pəreitid]adj. 组成公司的,合成一体的,合并的

unsaturated['ʌn'sætʃəreitid]adj. 没有饱和的,不饱和的

simplicity[sim'plisiti]n. 简单,简单性,单一性

chamber['tʃeimbə]n. 房间,会所

intermesh[ˌintə(:)'meʃ]vt. 使互相结合,使互相啮合

blade[bleid]n. 刀口,刀刃;叶片,草叶

banbury['bænbəri]n. 密炼机

nip[nip]n. 夹缝

velocity[vi'lɔsiti]n. 速度,速率

precaution[pri'kɔ:ʃən]n. 预防措施

systematic[ˌsistə'mætik]adj. 系统的,有体系的

2. Phrases and expressions

unsaturated polyester　不饱和聚酯

tumbling-type　滚动式

shear rate　剪切速率

power input　功率输入

banbury mixer　密炼机

two-roll mill　双辊塑炼机,双辊开炼机

3. Notes to the text

① Because of their extremely high melt viscosities, specialized equipment is usually needed to compound ingredients with high molecular weight thermoplastics, however. 由于物料的熔体黏度极高,混合高相对分子质量的热塑性塑料组分时常常需要使用专用设备。

② The band is cut off the roll with a knife, rolled up, and fed back to the nip at right angles to its former direction.　用刀把这片聚合物切下并卷起来,然后以正确的角度将其放回到辊隙中。

Lesson 43 Rubber Mixing

Mixing, being the first step in a sequence of operations, determines the efficiency with which subsequent processes may be carried out and exerts a considerable influence on product performance. Adequate and consistent mixing is a prerequisite for successful manufacture.

The type of system used for the mixing of rubber depends on both the form of the raw materials and the scale of the operation. The majority of rubber is supplied in bale form, dictating the use of batch mixing. In medium-and large-scale mixing, systems based on the internal mixer are used throughout the industry, while small-scale mixing is usually carried out on a two-roll mill.

The mixing of rubber is a complex operation and is very difficult to quantify. This complexity is a result of the viscoelastic behavior of the rubber and of the nature of the materials with which it is required to be mixed. Particulate fillers are not masses of simple particles but consist of groups of particles called agglomerates, which must be broken down and uniformly distributed throughout the rubber during mixing. Carbon blacks are pelletized, making the breakdown more difficult, and also consist of primary aggregates-clusters of particles which survive the mixing operation and influence the behavior of the finished product.

Liquids, in the form of oils, waxes, and plasticizers, are also added to rubber mixes and generate their own problems for mixer operation.

This text is primarily concerned with the internal mixer. However, the principles of mixing rubber with particulate materials and liquids are similar for all rubber mixing systems and common stages in the conversion of the raw material to a finished mix can be identified①. In addition, most mixing systems consist of an internal mixer and another machine-either a two-roll mill or an extruder/continuous mixer. The whole system must then be considered when setting the conditions necessary to produce an adequately mixed material.

Outline specifications of internal mixers from two major manufacturers of these machines are given in Tables 1 and 2. The smallest machines in each range are laboratory models, used mainly for compound development and mixing of test batches. General-rubber-goods manufacturers usually have mixers with chamber volumes in the range 40 ~ 250 liters while tire manufacturers, due to their larger volume requirements, tend to use machines in the range 250 ~ 700 liters②. It should be noted in Tables 1 and 2 (be omitted) that a chamber volume is given for the Farrel Bridge machines; but the volume of material which can be mixed is specified for the Francis Shaw machines, which is

60% ~75% of the chamber volume.

1. New words

prerequisite['pri:'rekwizit]n. 先决条件; adj. 首先具备的
scale[skeil]n. 规模
bale[beil]n. 团,块,包
batch[bætʃ]n. 一批,一次或一批的生产量
cluster['klʌstə]n. 颗粒团
aggregate['ægrigit]n. 团粒
pelletize['pelitaiz]vt. 制成颗粒,造粒

2. Phrases and expressions

internal mixer　密炼机

3. Notes to the text

① However, the principles of mixing rubber with particulate materials and liquids are similar for all rubber mixing systems and common stages in the conversion of the raw material to a finished mix can be identified.　但是,对所有橡胶混炼系统来说,橡胶与颗粒及液体物料的混炼原理都是相似的,从生胶转变至混炼胶都经历了相同的阶段。

② General-rubber-goods manufacturers usually have mixers with chamber volumes in the range 40 ~250 liters while tire manufacturers, due to their larger volume requirements, tend to use machines in the range 250 ~700 liters.　普通橡胶制品生产厂家一般使用混炼室容量为40 ~250L范围内的密炼机。而轮胎制造厂由于胶料需求量大,所以多使用混炼室容量在250 ~700L范围的密炼机。

4. Exercises

(1) What is the prerequisite for successful manufacture of rubber?

(2) Which equipment may be used better for the large-scale mixing, internal mixer or two-roll mill?

(3) Why can the clusters of particles in rubber influence the behavior of the finished product?

(4) Translate the last paragraph into Chinese.

(5) Use a two-roll mill available in laboratory to operate the machine for rubber mixing according to the procedure.

Reading Material

The Mechanisms of Rubber Mixing

The mixing of rubber is a composite operation, involving a number of different mechanisms and stages. These can be resolved into four basic processes:

(1) Viscosity reduction.

(2) Incorporation.

(3) Distributive mixing.

(4) Dispersive mixing.

Each of these can occur simultaneously and each can be the main rate-determining process, which will control the mixing time. The mixing time will depend on the type of compound being mixed and the mixing conditions.

When a charge of highly elastic rubber is fed into a mixer it must be rapidly converted to a state in which it will accept particulate additives. This stage is called viscosity reduction and is achieved by three interdependent mechanisms: temperature rise, chain extension, and mastication. Because of rubber's high viscosity and elastic stiffness, the initial deformation of the rubber in a mixer requires considerable mechanical energy, which is converted to heat causing a rapid temperature rise and viscosity reduction. In some rubbers, notably natural rubber, the viscosity is irreversibly changed by chain scission, whereas the changes due to a rise in temperature and chain extension are recoverable. Consequently, viscosity reduction from chain scission (mastication) influences mixed-compound behavior and must therefore be controlled. Despite adding yet another variable to the mixing process, mastication provides the opportunity of achieving uniform mixed-material properties from a variable feedstock.

As the viscosity and elasticity of a rubber are reduced the rubber can be caused to flow around additives, incorporating and enclosing them in a matrix of rubber. The efficiency of incorporation is dependent upon free-surface "folding" flows being induced in the rubber by the mixer, overlapping and enclosing volumes of additives①.

Incorporation and distributive mixing generally proceed simultaneously, the latter commencing as soon as incorporated additives are available for distribution. These two types of mixing are accompanied by subdivision, in which the size of the volumes of additives is progressively reduced. Incorporation, subdivision, and distribution are largely due to exponential mixing mechanisms.

During incorporation, subdivision, and distributive mixing the rubber flows around filler-particle agglomerates and penetrates the interstices between particles in the agglomerates. This action has two effects. First, due to the "wetting out" of the filler by the rubber and reduction of voids, the rubber mix becomes less compressible and its density increases. Second, the rubber which has penetrated the interstices becomes immobilized and is no longer available for flow②. Medalia refers to this as occluded rubber and points out that immobilization reduces the effective rubber content of the mixture. This reduction has been effect of increasing viscosity; the incompressibility of the mixture now allows high forces to be applied to the particle agglomerates, causing them to fracture. This action is termed dispersive mixing, and will continue while the forces being applied to particle agglomerates, of both fillers and minor additives, are sufficient to cause fracture.

Distributive mixing occurs concurrently with dispersive mixing, which serves the purpose of separating the fragments of agglomerates once they have been fractured.

Many particulate additives are pelletized or produced in flakes and other forms which give rapid incorporation. In the absence of a mastication stage, mixing time is generally dictated by distributive mixing when large particle size diluent or semireinforcing fillers are used; and by dispersive mixing when reinforcing fillers, particularly carbon blacks, are used③. Mohr also points out that the smaller the volume fraction of a minor additive, the more mixing is needed to ensure a uniform distribution of that additive. It is more difficult to mix a small amount into a large amount than it is to achieve an acceptable 50-50 mixture.

1. New words

interdependent[ˌintəːdiˈpendənt] *adj.* 相互关联的
notably[ˈnəutəbli] *adv.* 尤其是，值得注意地
irreversibly[ˌiriˈvəːsəbli] *adv.* 不可逆地，单向地
feedstock[ˈfiːdstɔk] *n.* 胶料，原料
enclose[inˈkləuz] *vt.* 包裹
matrix[ˈmætriks] *n.* 基体
folding[ˈfəuldiŋ] *n.* 折叠
overlapping[ˌəuvəˈlæpiŋ] *n.* 复叠，重叠
commence[kəˈmens] *vt.* 开始，着手
subdivision[ˌsʌbdiˈviʒən] *n.* 细分，分段
progressively[prəˈgresivli] *adv.* 逐渐地
interstice[inˈtəːstis] *n.* 间隙，空隙
void[ˈvɔid] *n.* 空隙，间隙
compressible[kəmˈpresəbl] *adj.* 可压缩的
immobilize[iˈməubilaiz] *vt.* 不能活动
fracture[ˈfræktʃə] *n.* 破裂，断裂
fragment[ˈfrægmənt] *n.* 碎片，碎屑
scission[ˈsiʒən] *n.* 剪断，分隔

2. Phrases and expressions

free surface 自由表面
wetting out 浸润
dispersive mixing 分散混炼
distributive mixing 分布混炼
large particle size diluent 粗粒增容剂
semireinforcing fillers 半补强填料

3. Notes to the text

① The efficiency of incorporation is dependent upon free-surface "folding" flows being induced in the rubber by the mixer, overlapping and enclosing volumes of additives. 混合效率取决于混炼机中胶料之间自由表面产生的折叠流动，以及复叠和包裹添加剂的流动。

② This action has two effects. First, due to the "wetting out" of the filler by the rubber and reduction of voids, the rubber mix becomes less compressible and its density increases. Second, the rubber which has penetrated the interstices becomes immobilized and is no longer available for flow.　此渗入有两个作用：第一，由于橡胶"浸润"了填料，填料间的空隙减少了，混合胶料的可压缩性减小了，因而它的相对密度增加了；第二，已渗入空隙中的橡胶变得不能活动，也就不可能再流动了。

③ In the absence of a mastication stage, mixing time is generally dictated by distributive mixing when large particle size diluent or semireinforcing fillers are used; and by dispersive mixing when reinforcing fillers, particularly carbon blacks, are used.　在没有捏合阶段，当使用粗粒增容剂或半补强填料时，其混炼时间一般取决于分布混炼；当使用补强填料，特别是炭黑时，其混炼时间则取决于分散混炼。

Lesson 44 Rubber Extrusion and Continuous Mixing

Extruders are widely used in the rubber industry in a variety of applications. In large mixing systems, dump extruders are used to accept the batch of material from an internal mixer and to give it a shape suitable for further operations. Again in the mixing system, mixing extruders or continuous mixers are used to incorporate and distribute particulate additives. Further down the production line, extruders are used to preform rubber for further operations and to form finished products. All these applications generate their own machine performance requirements, and the wide range of extruder designs available reflects this.

Extruders may be categorized in two ways. First, extruders may be identified by the temperature of the feedstock necessary for successful operation. Traditionally, hot-feed extruders have been used by the rubber industry, where the feedstock is prewarmed in a prior operation. For conventional hot-feed extrusion a two-roll mill is usually used for prewarming. Cold-feed extruders, taking strip or granulated rubber at ambient workshop temperature are a more recent introduction, probably resulting from the advances in extruder design for the plastics industry. Second, extruders may be identified by application. Many companies require an "undedicated" machine which is capable of operating successfully, if not efficiently, with a wide range of rubber mix types. Here the emphasis in design is to minimize the time taken to change a die and return the extruder to useful operating conditions, and to achieve efficient self-purging to minimize cross-contamination from mix changes. When an extruder is to be used for long runs with rubber mixes having a narrow range of flow properties, the screw, head, and die can be designed to give both high output rates and good dimensional control. Also, the feed and haul-off equipment and the control system may be selected to ensure that the good dimensional control is maintained, despite minor variations in the feed material.

The major physical difference between hot-and cold-feed extruders lies in the length-to-diameter ratio of the screw. For hot-feed machines, where a considerable portion of the input of energy to the rubber mix for heating and preplasticizing is carried out on a two-roll mill, the functions of the extruder screw are simply those of conveying and pressurizing①. This has resulted in "short" machines having screw lengths, in terms of their diameters, of $3D$ to $5D$. In addition to conveying and pressurizing, the screw of a cold-feed extruder must input to the rubber all the mechanical work necessary to raise it to the desired temperature for smooth flow through the die. This requires screws having lengths in the region of $9D$ to $15D$, and for some applications longer screws than these may be used.

Cold-feed extruders have largely replaced hot-feed types in production lines where long runs are achieved and where good dimensional accuracy is required, and have made considerable inroads into the "undedicated" area with improvements in versatility resulting from design development and operating "know-how"②. However, hot-feed extruders are widely used, and Iddon points out that the capital cost and energy consumption of conventional cold-feed extruders increase rapidly when the screw diameter exceeds approximately 150mm.

1. New words

incorporate[in'kɔpəreit] *vi.* 混合
categorize['kætigəraiz] *vt.* 分类
prewarm['pri:'wɔ:m] *v.* 预热
ambient['æmbiənt] *adj.* 周围的
undedicated['ʌn'dedikeit] *adj.* 非专用的
inroad['inrəud] *n.* 改进

2. Phrases and expressions

dump extruder 卸料挤出机
ambient workshop temperature 室温
self-purging 自行清洗
cross-contamination 交叉污染
length-diameter ratio 长径比
know-how 专业知识

3. Notes to the text

① For hot-feed machines, where a considerable portion of the input of energy to the rubber mix for heating and preplasticizing is carried out on a two-roll mill, the functions of the extruder screw are simply those of conveying and pressurizing. 对热喂料挤出机来说,其输给混炼胶的大部分能量是用来在两辊开炼机上进行加热和塑化胶料,而挤出机螺杆的作用仅是输送和压缩胶料。

② Cold-feed extruders have largely replaced hot-feed types in production lines where long runs are achieved and where good dimensional accuracy is required, and have made considerable inroads into the "undedicated" area with improvements in versatility resulting from design development and operation "know-how." 在进行批量生产和对产品尺寸精度要求高的生产线上,冷喂料挤出机已经取代了热喂料挤出机,而且由于设计上的改进和专业知识的增加,冷喂料挤出机的通用性更强了,在"非专用"领域已经独占鳌头。

4. Exercises

(1) What are the extruders used for in the rubber industry?

(2) How can extruders be categorized and what are they?

(3) What are the major physical differences between hot-and cold-feed extruders?

(4) Why have hot-feed extruders been replaced by cold-feed types recently?

(5) The length-to-diameter ratio of the screw for hot-feed extruder is ________ to ________ and for cold-feed extruder is ________ to ________.

(6) Make a drawing of the extruder in your laboratory showing the extruder screw,

cylinder and die in cross section. Identify the difference between a plastic extruder and a rubber extruder.

Reading Material

Rubber Calendering and Vulcanization

Calendering: Calendering and milling are sufficiently similar to enable them to be treated together. Both have been used for many years in the rubber industry, resulting in a considerable fund of practical expertise for their operation. However, technical studies of both the two-roll mill and calender yield information which can be put to practical use in any company.

Residence time may be used to differentiate conveniently between calendering and milling. Two-roll milling involves a substantial residence time and many passes through a single rolling nip. This is necessary for mixing or for raising the temperature of premixed material to that required by subsequent processes. Calendering is essentially a shaping operation where the work done on the material is required to produce a change of shape, not a change of state[①]. This involves only a small number of passes between rolling nips, but does require prewarming and homogenization of the feed material-on a two-roll mill, for example. For continuity of production, calendering operations which use more than one rolling nip have to be performed on machines with the requisite number of rolls to form these nips. For effective mixing and homogenization in a rolling nip, a bank of material must be formed above the nip. This reservoir of material ensures that the nip is adequately fed and that effective flow work is being done on the material.

Vulcanization: Vulcanization processes divide naturally into two main groups. The first consists of molding methods, all of which involve an integral shaping operation which is completed prior to the onset of cross-linking. The second includes a number of techniques used to cure a previously formed product. For the purposes of analysis, the shaping operations in molding can be considered to be separate from the vulcanization stage, enabling the majority of vulcanization processes to be evaluated using similar techniques.

During vulcanization externally supplied heat flows into the rubber at a rate controlled by the efficiency of heat transfer from the heating medium and by the heat-transfer properties of the rubber[②]. The temperature gradients in the rubber arising from this conductive heat transfer then depend on the temperature of the external heat source, the time of heating, the size and shape of the article being vulcanized, and its initial temperature. Changes in temperature within a rubber product, which occur with respect to both time and position, tend to give a nonuniform state of cure and can result in the properties of the rubber at the surface of a product being quite different from those at the center. One of the main objectives of selection and optimization of vulcanization processes is that of achieving an acceptably uniform state of cure in conjunction with a viable production rate.

Heat transfer within a rubber product during vulcanization is conductive in nature, except in the case of radiation curing processes. However, the mode of supply of heat to the surface of the article can be either conductive or convective.

1. New words

homogenization [həˌmɔdʒənaiˈzeiʃən] *n.* 均匀化

reservoir [ˈrezəvwaː] *n.* 堆积,储藏

convective [kənˈvektiv] *adj.* 对流的

2. Phrases and expressions

residence time 停留时间

rolling nip 辊距

flow work 流动功

in conjunction with 在……同时

radiation curing 辐射硫化

3. Notes to the text

① Calendering is essentially a shaping operation where the work done on the material is required to produce a change of shape, not a change of state. 压延主要是一种成型作业,它对胶料所要求进行的加工是改变它的形状,而不改变它的物性。

② During vulcanization externally supplied heat flows into the rubber at a rate controlled by the efficiency of heat transfer from the heating medium and by the heat-transfer properties of the rubber. 在硫化过程中,外部供给的热量传入橡胶的速率是由载热体的传热效率和橡胶的传热性质所控制的。

Lesson 45 Fiber Spinning of Polymer

The first efforts of manufacturing artificial fibers were connected with the attempts to utilize natural polymers, primarily cellulose and casein, as raw materials. Although synthetic polymers were already known at that time, they were considered merely as curiosa without utilitarian value. Macromolecular chemistry as a science was to be born much later. Therefore, natural products similar to known natural fibers were much more trustworthy. The natural polymeric raw materials used to manufacture the first artificial fibers did not melt at elevated temperatures without decomposition. This fact determined the first technological approaches to fiber making: namely, in order to change the physical shape of the raw material, it had to be first dissolved, a solution formed, and the solvent subsequently removed. In such a way the methods of fiber formation from solution were first conceived. The remaining questions were: how to dissolve the polymer, and how to get the polymer back into the solid state after it had been given the desired shape.

Since cellulose does not dissolve easily, three different approaches were originally implemented:

(1) Cellulose was dissolved in cuprammonium.

(2) Cellulose was chemically modified to nitrocellulose to make it soluble.

(3) Cellulose was modified chemically to assure solubility of the material, but immediately after shaping it, the transition compound, cellulose xanthate, was decomposed back to cellulose①.

Each of the methods predetermined, to some extent, the possible methods of solidification. In the cases of methods 1 and 2 the solvent may be either evaporated or extracted. The third case requires a chemical reaction to take place, so a wet process, similar to that involving extraction, was necessary.

Much later, fully synthetic polymers were considered as raw materials for fiber manufacturing and these polymers had thermoplastic properties. This fortunate circumstance allowed the omission of the processing auxiliary in the form of solvent, and so fiber formation direct from the polymer melt came into being②.

Traditionally, the three methods of fiber formation were treated as entirely different processes having only very little in common. From an operational point of view, this opinion is justified to only some extent. From theoretical and general technological points of view, all of the processes are very similar.

Figure 36. 1 shows a general scheme of spinning with all three of the main systems. In the first stage the polymer is either dissolved or melted. Dissolution is performed mostly in batches, while melting, in the majority of contemporary processes, is carried out continuously in screw melters. Occasionally, one may still meet melt spinning process with other feeding systems, including batch feeding of the melt under gas pressure or feeding of the granular polymer which is melted on grill heaters just before entering the spinning block. In both cases, polymer, as solution or melt, is transported under pressure to spinning blocks where an exact metering pump, e. g. a gear pump, maintains a highly even issue of polymer[③]. After passing the metering pump, the polymer, either as melt or as solution, is forced through the final filter usually called the filtering pack, which has more functions than just mechanical purification. Finally, the liquid raw material is forced through a plate with capillaries, called the spinnerette, and in this way is formed in endless, fine streams of liquid. The spinning blocks for melt and dry spinning are of a similar design principle, except the higher melt viscosities require higher strength and heavier construction. The spinnerettes for wet spinning are usually mounted not directly in the spinning block but at the end of special transfer tubes attached to the block. Such a design change, in comparison with melt and dry spinning, is more convenient, since the spinnerettes must be submerged in a liquid(Figure 45. 1).

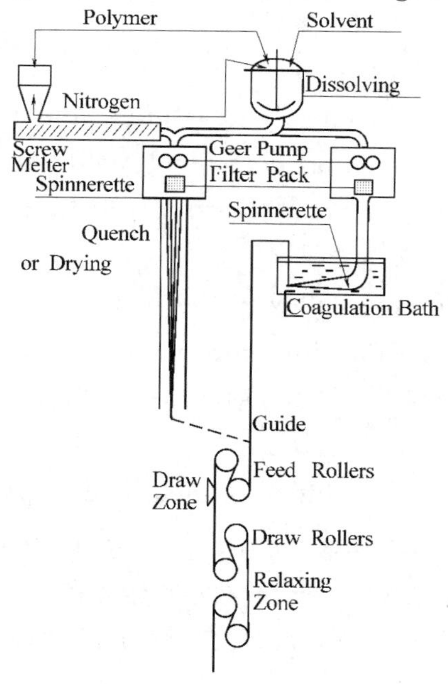

Figure 45. 1 Schematic representation of all spinning systems

1. New words

spinning['spiniŋ] *n.* 纺织,纺纱

curiosa[ˌkjuəri'əusə] *n.* 珍品,奇品

utilitarian[ˌju:tili'tεəriən] *adj.* 有效用的,实用的

elevated['eliveitid] *adj.* 提高的,升高的

cuprammonium[ˌkju:prə'məuniəm] *n.* 四氨络铜离子,铜铵离子,铜铵液

nitrocellulose[naitrəu'seljuləus] *n.* 硝化纤维,硝酸纤维素

xanthate['zænθeit] *n.* 黄酸盐,黄原酸盐,

黄原酸酯

predetermine['pri:di'tə:min] *vt.* 预定,预先确定

extract[iks'trækt] *vt.* & *n.* 提取,榨出,引用,提炼物,浓缩物

omission[əu'miʃən] *n.* 省略,遗漏,疏忽

granular['grænjulə] *adj.* 粒状的,颗粒状的

grill[gril] *n.* 烤架,烤肉;*v.* 烧,烤

filter['filtə] *n.* 滤波器,过滤器,滤色镜;*v.* 过滤,渗透

purification[ˌpjuərifi'keiʃən] *n.* 净化,提纯

capillary[kə'piləri] *n.* 毛细管,毛细血管

submerge[səb'mə:dʒ] *vt.* &*vi.* (使)潜入水中,淹没,完全掩盖,遮掩

2. Phrases and expressions

artificial fibers　人造纤维

technological approaches　技术途径

cellulose xanthate　黄酸纤维素

wet process　湿法加工,湿法

in batches　分批地,成批地,批量地

batch feeding　投料,配合料加料

granular polymer　粒状聚合物

spinning block　旋压模

metering pump　计量泵

melt spinning　熔融纺丝

dry spinning　干式纺丝,(亚麻)干纺

wet spinning　湿纺,湿法纺丝

3. Notes to the text

① Cellulose was modified chemically to assure solubility of the material, but immediately after shaping it, the transition compound, cellulose xanthate, was decomposed back to cellulose.　使纤维素化学变性以保证物料的可溶解性,但在成型后立即使过渡化合物(即黄酸纤维素)分解,重新回复为纤维素。

② This fortunate circumstance allowed the omission of the processing auxiliary in the form of solvent, and so fiber formation direct from the polymer melt came into being.　这种有利的性质可以省去溶剂一类的加工助剂,因此出现了由聚合物熔体直接成型为纤维的方法。

③ In both cases, polymer, as solution or melt, is transported under pressure to spinning blocks where an exact metering pump, e. g. a gear pump, maintains a highly even issue of polymer.　在这两种情况下,聚合物溶液或熔体在压力下输入纺织组件,在这个组件中,由一个精密计量泵,例如齿轮泵,保持聚合物流量的高度均衡。

4. Exercises

(1) When and how did artificial fibers come into our life?

(2) What's the main problem with natural polymeric raw materials and what's the subsequent result?

(3) What are the remaining questions and how do you solve them?

(4) What's the function of spinning block in a spinning system?

(5) What're the differences between wet spinning and melt or dry spinning in prin-

ciple, design and application?

(6) Translate the following sentence into Chinese:

Occasionally, one may still meet melt spinning process with other feeding systems, including batch feeding of the melt under gas pressure or feeding of the granular polymer which is melted on grill heaters just before entering the spinning block.

Reading Material

Principles of the Melt-Spinning of Polymer

The melt-spinning procedure consists of the preparation of the spinnable polymer melt, extrusion of the melt through spinneret orifices, extension of polymer jets leaving the orifices, and the winding up of the solidified filaments on a bobbin or a similar take-up element.

In the case of multifilament yarns and staples cooking of the spinning line proceeds most often in a gaseous medium (air, steam) in a path 1-5 m long. The cooling effect is sometimes increased through a perpendicular gas flow which is also designed to improve the stability of the spinning process. On the other hand, in viscoelastic materials with low fluidity, heated chambers below the spineret are used to prevent too rapid a cooling. Thick monofilaments (bristles) are usually spun into rapidly acting liquid cooling baths.

The spinning (takeup) velocities used in fiber formation from polymer melts range from 100 m/min (thick monofilaments in liquid baths) to several thousand meters per minute (multifilaments in a gaseous medium). The lack of solvents and precipitating agents and high velocities employed apparently make melt-spinning the most convenient and efficient of spinning procedures. The only factor seriously limiting its application is connected with the polymer material. Fiber-forming polymer to be melt-spun must yield stable fluid melts in technically available temperatures (rather below 300℃)[①]. Typical melt-spun polymers are linear polycondensates (polyamides, polyesters, polyurethanes) and crystalline polyolefins (polyethylene, polypropylene). Thermal destruction of a polymer below the fluidity temperature (exhibited, for example, by cellulose and polyvinyl alcohol) excludes the application of the melt-spinning procedure. This difficulty, however, is sometimes circumvented by replacing pure, undiluted polymers by plasticized systems, i. e., polymers containing a few per cent of low-molecular plasticizer decreasing the fluidity temperature below that of destruction, thus making extrusion possible[②]. An example of such a process is provided by polyacrylonitrile. Although in this case the spinning fluid is actually a very concentrated solution, such a process should be considered as "melt-spinning" rather than "solution dry-spinning" because, as in pure melts, the mechanism of solidification consists of cooling, rather than of evaporation of solvent (plasticizer).

Melt-spinning is usually followed by the mechanical treatment of solidified fibers (drawing), leading to molecular orientation along the fiber axis and to improving of the physical characteristics of fibers. This operation, consisting in two-to sixfold elongation of melt-spun fibers, may be accomplished directly after spinning (continuous drawing) or separately, starting from undrawn material.

The general fundamentals of the fiber-spinning process have already been discussed. There are many variables involved in melt-spinning which determine the course of fiber formation and the resulting fiber's dimensions and properties. Some of the variables are mutually dependent; under steady-state conditions the equation of continuity holds:

$$\rho_0 A_0 V_0 = \rho_L A_L V_L = W \quad (45.1)$$

where ρ_0, ρ_L are the polymer densities, A_0, A_L are the cross-sectional areas of the spinning line, V_0, V_L is the average velocity of the spinning line at the spinneret exit ($x = 0$) and at the takeup device ($x = L$), respectively, and W is the mass output rate.

All the spinning variables can be divided into three groups:

(1) Independent, or primary, which uniquely determine the course of the spinning process and the resulting fiber structure and properties.

(2) Secondary, related to primary ones through equation 1, alternatively used for describing the spinning conditions.

(3) Resulting, determined by independent variables and by the fundamental laws of spinning kinetics③.

1. New words

multifilament[ˌmʌlti'filəmənt] *n.* 复丝,多纤(维)丝

monofilament['mɔnəu'filəmənt] *n.* 单(根长)丝,单纤(维)丝

yarn[ja:n] *n.* 纱,纱线

jet[dʒet] *n.* 喷流,射流

robbin['rɔbin] *n.* 筒,管

spinneret[ˌspinə'ret] *n.* 喷丝头,纺丝头

bristle['brisl] *n.* 硬毛,鬃

takeup['teik'ʌp] *n.* 卷绕

polyurethanes['pɔli'juəriθein] *n.* 聚氨酯,聚氨基甲酸酯

2. Phrases and expressions

precipitating agent 沉淀剂

takeup velocity 引出速度,卷取速度

3. Notes to the text

① Fiber-forming polymer to be melt-spun must yield stable fluid melts in technically available temperatures(rather below 300℃). 用来做熔融纺丝的成纤高聚物必须是在技术条件能够达到的温度(大大低于 300℃)下的稳定流动熔体。

② This difficulty, however, is sometimes circumvented by replacing pure, undiluted polymers by plasticized systems, i. e. , polymers containing a few per cent of low-molecular plasticizer decreasing the fluidity temperature below that of destruction, thus making extrusion possible. 然而这个困难有时是可以克服的,例如在聚合物中加入百分之几的低分子增塑剂以代替未稀释的纯聚合物,使聚合物的流动温度降低到分解温度以下,从而使挤出成为可能。

③ Resulting, determined by independent variables and by the fundamental laws of spinning kinetics. 终结变数:这是由独立的变数和纺丝动力学的基本定律所决定的变数。

Glossary
总词汇表

abrasion[ə'breiʒən] *n.* 磨损,磨蚀
accelerator[æk'seləreitə] *n.* 加速器
accumulate[ə'kju:mjuleit] *vi.* & *vt.* 聚积积累
acetal['æsitæl] *n.* 聚甲醛,缩醛
acetate['æsitit] *n.* 醋酸盐
acetone[' 'æsitəun] *n.* 丙酮
acetylene[ə'setli:n] *n.* 乙炔
acrylic[ə'krilik] *adj.* 丙烯酸的
adapter[ə'dæptə] *n.* 连接器
addition[ə'diʃən] *n.* 加成作用
additive['æditiv] *n.* 添加剂,助剂
adherend[əd'hiərənd] *n.* (使用黏合剂的)黏着面,被黏物(黏附体)
adhesive[əd'hi:siv] *n.* 黏合剂; *adj.* 有黏性的
adhesive[əd'hi:siv] *n.* 黏合剂,胶黏剂; *adj.* 可黏着的,黏性的
adjustable[ə'dʒʌstəlb] *adj.* 可调节的
affinity[ə'finiti] *n.* 亲合力,亲和力
agglomerate[ə'glɔməreit] *n.* 大团,大块
aggregate['ægrigit] *n.* 团粒
agitate['ædʒiteit] *vt.* 搅拌
aircraft['ɛəkra:ft] *n.* 航空器
akin[ə'kin] *adj.* 酷似,类似的
albeit[ɔ:l'bi:t] *conj.* 尽管,即使
alchemy['ælkəmi] *n.* 炼丹术
alcohol['ælkəhɔl] *n.* 酒精,乙醇
align[ə'lain] *vi.* 排列,排成一行
aliphatic[ˌæli'fætik] *adj.* 脂肪族的
alkyd['ælkid] *n.* 醇酸
alkyl['ælkil] *n.* 烷(烃)基
aluminum[ˌælju'miniəm] *n.* 铝
amber['æmbə] *n.* 琥珀
ambient['æmbiənt] *adj.* 周围的
amide['æmaid] *n.* 酰胺
amine['æmi:n] *n.* 胺
amino['æminəu] *adj.* 氨基的
amorphous[ə'mɔ:fəs] *adj.* 无定形的,无组织的,[物]非晶形的
analogous[ə'næləgəs] *adj.* 类似的,相似的
angular['æŋgjulə] *adj.* 有角的,角度的
anion['ænaiən] *n.* 阴离子,负离子
anionic['ænaiənik] *adj.* 阴离子的,具有活性阴离子的
anomalous[ə'nɔmələs] *adj.* 反常的,不规则的
antioxidant[ænti'ɔksid(ə)nt] *n.* [助剂]抗氧化剂
antioxidant['ænti'ɔksidənt] *n.* 防老剂
apparatus[ˌæpə'reitəs] *n.* 装置,设备,仪器,器官
appreciation[əˌpri:ʃi'eiʃən] *n.* 理解,领会
approximation[ə'prɔksimeiʃən] *n.* 近似值
aqueous['eikwiəs] *adj.* 含水的,水的
arable['ærəbl] *adj.* 可耕的,可开垦的
arbitrarily['a:bitrərili] *adv.* 任性地,专断地
arithmetic[ə'riθmətik] *n.* 算术,算法
armature['ɑ:mətʃə] *n.* 电枢(电机的部件),盔甲
aromatic[ˌærəu'mætik] *adj.* 芳香的,醇香的,芬芳的
article['a:tikl] *n.* 制品
artifact['ɑ:tifækt] *n.* 人造物品
artificial[a:ti'fiʃəl] *adj.* 人造的,模拟的
asphalt['æsfælt] *n.* 沥青,柏油
astonish[əs'tɔniʃ] *v.* 使惊讶,惊奇
asymmetric[ˌæsi'metrIk] *adj.* 不对称的,

非对称的

asymptotically[əˌsaimp'tɔtikəli]adv. 渐近

atactic[e'tæktik]adj. 不规则的,[有化]无规立构的

autoacceleration[ækˌsel'əreiʃən]n. 自动加速效应

auxiliary[ɔ:g'ziljəri]adj. 辅助的,副的

availability[əveilə'biliti]n. 适用性,使用价值

azo['æzəu]adj. 含氮的

azobisisobutyronitrile (AIBN)[ˌæzəubis'aisouˌbju:ti'rɔnaitrail]n. 偶氮二异丁腈

backbone['bækbəun]n. 主链

bakelite['beikəlait]n. 电木,胶木

bale[beil]n. 团,块,包

banbury['bænbəri]n. 密炼机

bare[bɛə]adj. & vt. 赤裸的,无遮蔽的,空的,使赤裸,露出

barrel['bærəl]n. 料筒,桶;vt. 装入桶内

batch[bætʃ]n. 一批,一次或一批的生产量

bead[bi:d]n. 珠粒

bearing['bɛəriŋ]n. 轴承

belated[bi'leitid]adj. 误期的,迟来的

bending['bendiŋ]n. 弯曲(度),挠度

benzoyl[benzəuil]n. 苯甲酰基

beryllium[bə'riljəm]n. 铍

beverage['bevəridʒ]n. 饮料

billiard['bilɪrd]adj. 台球的

bimodal[bai'məud(ə)l]adj. 双峰的

binder['baində]n. 黏合剂

binocular[bi'nɔkjulə]adj. [生物]双眼的,双目并用的;n. 双筒望远镜

birefringent[ˌbairi'frindʒənt]adj. [光]双折射的

blade[bleid]n. 刀口,刀刃;叶片,草叶

blending['blendiŋ]n. 共混

bolt[bəult]vt. & vi. 用螺栓栓上

bond[bɔnd]vi. 粘接

botulism['bɔtjulizəm]n. 肉毒中毒(食物中毒的一种)

bouncing['baunsiŋ]adj. 跳跃的,活泼的,巨大的

branch[bræntʃ]n. 分枝(支链)

branched[bræntʃd]adj. 支化的

breakage['breikidʒ]n. 裂口

brew[bru:]n. 酿造,煎药

bristle['brisl]n. 硬毛,鬃

brittle['britl]adj. 易碎的

bubble['bʌbl]n. 膜泡

built-in[ˌbilt'in]adj. 内置的,固定的,嵌入的;n. 内置

bulk[bʌlk]n. 大小,体积,大批;vt. 显得大,显得重要;adj. 大批合计的

bumper['bʌmpə]n. 缓冲器

butadiene[ˌbju:tə'daɪi:n]n. 丁二烯

butyllithium['bju:til'liθiəm]n. 丁基锂

butyrate['bju:tireit]n. 丁基盐(酯)

calender['kælində]v. 压延

calorimetry[ˌkælə'rimitri]n. [热]量热学,热量测定

camphor['kæmfə]n. 樟脑

canvas['kænvəs]n. 帆布

capillary[kə'piləri]n. 毛细管,毛细血管

capstan['kæpstən]n. 绞盘,起锚机

carbolic['ka:bɔlik]adj. 碳的

carboxyl[ka:'bɔksil]n. 羧基

carpet['ka:pit]n. 地毯

casein['keisi:in]n. 干酪素,酪蛋白

casting['ka:stiŋ]n. 铸塑,流涎

catalyst['kætəlist]n. 催化剂

categorize['kætigəraiz]vt. 分类

category['kætigəri]n. 种类,类别

cation['kætaiən]n. 阳离子,正离子

cavity['kæviti]n. 洞,空穴,腔

cell[sel]n. 泡孔

cellular['seljulə]adj. 多孔的,泡沫的

celluloid['seljulɔid]n. 赛璐珞

cellulose['seljuləus]n. 纤维素

cellulosic[ˌselju'ləusik]adj. 纤维素的,有

纤维质的
cement[si'ment]n. 黏合剂
chamber['tʃeimbə]n. 房间，会所
characteristic [ˌkæriktə'ristik] n. 特征；adj. 特有的，典型的，特性
characterize['kæriktəraiz]vt. 表示……的特色，赋予……的特色
charring[tʃɑ:riŋ]n. 碳化，烧焦
cheddar['tʃedə]n. 干酪的一种
chloride['klɔ:raid]n. 氯化物
chlorinated['klɔ:rineitid]adj. 氯化的
chlorinated['klɔrinetid]adj. 含氯的
chloroprene['klɔ:rəpri:n]n. 氯丁二烯
chromatography['krəumə'tɔgrəfi]n. 色层分析，色谱分析法
chronologically[ˌkrɔnə'lɔdʒikli]adv. 按时间的前后顺序排列地
clammy['klæmi]adj. 滑腻的，黏的
clamp[klæmp]vt. 合模
clarity['klæriti]n. 透明，清晰度
classify['klæsifai]vt. 分类，归类
clay[klei]n. 黏土
cling[kliŋ]vi. 黏着，依附于
clothesline['kləuðzlain]n. 晾衣绳
cluster['klʌstə]n. 颗粒团
coagulation[kəuˌægju'leiʃən]n. 凝结
coalesce[kəuə'les]vi. 联合，合并
cocatalyst[kəu'kætəlist]n. 助催化剂
cohesive[kəu'hi:siv]adj. 有结合力的，产生结合力的，产生内聚力的
coil[kɔil]vt. 卷取
coin[kɔin]vt. 制造
collapse[kə'læps]v. 压扁[平]，毁坏，断裂
colloidal[kə'lɔidl]adj. 胶状的，胶质的
Columbus[kə'lʌmbəs]n. 哥伦布
combination[kɔmbi'neiʃən]n. 组成
commence[kə'mens]vt. 开始，着手
commensurate[kə'menʃərit]adj. 相应的，相称的
component[kəm'pəunənt]adj. 组成的，合成的，成分的，分量的
compounder[kɔm'paundə]n. 混炼机
compounding [ˌkɔmpaundiŋ] n. 组合，混合
compressible[kəm'presəbl]adj. 可压缩的
compression[kəm'preʃən]n. 挤压，压缩
compromise['kɔmprəmaiz]n. & v. 妥协，和解，兼顾
concentrate['kɔnsentreit]n. 浓缩物，母料
concrete ['kɔnkri:t] n. 水泥，混凝土；adj. ；具体的，实在的；vi. 凝结，结合
condensation [kɔnden'seiʃən] n. 压缩，缩聚
configuration[kənˌfigjə'reʃən]n. 构型，配置，结构，外形
configuration[kənˌfigju'reiʃən]n. 构型
configurational [kənˌfigju'reiʃən] adj. 构造的
confined[kən'faind] adj. 限于，被限制的，狭窄的
conformation[ˌkɑnfɔr'meʃən]n. 构造，一致，符合，构象
consistent[kən'sistənt]adj. 相等的
console[kən'səul]n. 控制台
constituent[kən'stitjuənt]n. 成分；adj. 构成的，组织的
consult[kən'sʌlt]v. 咨询
contemporary [kən'tempərəri] adj. 当代的，同时代的
contour['kɔntuə]n. 外形，轮廓
contract[kən'trækt]v. 收缩，缩短
controversy['kɔntrəvə:si]n. 争议，论争，辩论
convective[kən'vektiv]adj. 对流的
conversion[kən'və:ʃən]n. 转化(率)
convert[kən'və:t]vt. 使转变
copolymerization [kəuˌpɔliməri'zeiʃən] n. 共聚
core[kɔ:]n. 模芯，阳模

corrosion[kə'rəuʒən]n. 腐蚀
corrosive[kə'rəusiv]adj. 腐蚀的,侵蚀性的
corrosive[kə'rəusiv]adj. 腐蚀性的,侵蚀性的
counterpart['kauntəpa:t]n. 对应物
counterproductive['kauntəprəˌdʌktiv]adj. 反生产的,使达不到预期目标的
couple['kʌpl]n. 连接,配合
covalent[kəu'veilənt]adj. 共有原子价的,共价的
cranny['kræni]n. 裂缝,裂隙
creep[kri:p]n. 蠕变
crevice['krevis]n. 裂缝,缺口
criterion[krai'tiəriən]n. 标准(*Pl.* -criteria)
croquet['krəukei]n. 槌球戏,循环球戏
crosshead['krɔshed]n. 十字头,丁字头,小标题,子题
crosslink['krɔsliŋk]vt. 交联;n. 交联点
crosslinked['krɔsliŋkd]adj. 交联的
crosslinking[krɔs'liŋkiŋ]n. 交联
crush[krʌʃ]vt. 压碎,压裂
crystal['kristəl]n. 水晶,结晶(体);adj. 水晶制的,水晶般的,透明的
crystalline['krist(ə)lain]adj. 透明的,结晶的,水晶制的;n. 结晶质
crystallinity['kristəlainiti]n. 结晶
crystallization['kristəlai'zeiʃən]n. 结晶化
crystallize['kristəlaiz]vt. 使……结晶
cuprammonium[ˌkju:prə'məuniəm]n. 四氨络铜离子,铜铵离子,铜铵液
cure[kjuə]n. 固化,硫化
curing['kjuəriŋ]n. 固化,熟化,硫化
curiosa[ˌkjuəri'əusə]n. 珍品,奇品
cushion['kuʃən]v. 缓冲
cyanoacrylate[ˌsaiənəu'ækrileit]n. 氰基丙烯酸盐黏合剂,氰丙烯酸酯
cylinder['stilində]n. 料筒
cylindrical[si'lindrikəl]adj. 圆柱形的,圆柱体的
dash[dæʃ]n. 仪表板(=dash panel)
dashpot['dæʃpɔt]n. 转折点,黏壶
decomposition[ˌdi:kɔmpə'ziʃən]n. 分解,腐烂
deep-drawn[ˌdi:p'drɔ:n]adj. 深长的
defective[di'fektiv]adj. 残次的
deform[di'fɔ:m]v. 变形
degradation['degrə'deiʃən]n. 退化,降解
degrade[di'greid]vt. & vi. (使)降解,(使)退化
deleterious[deli'tiəriəs]adj. 有害的,有害杂质的
delicate['delikit]adj. 精致的
denote[di'nəut]n. 表示,代表
density['densiti]n. 密度
detergent[di'tə:dʒənt]adj. 清洁的;n. 清洁剂
deterioration[diˌtiəriə'reiʃən]n. 恶化
deterrent[di'terənt]n. 阻碍物
deviation[ˌdi:vi'eiʃən]n. 偏差,偏离
devulcanized[di:'vʌlkənaizd]adj. 脱硫的
diameter[dai'æmitə]n. 直径
diametrically[ˌdaiə'metrikəli]adv. 完全地
diamine[daiə'mi:n]n. 二胺化合物
dianion[dai'ænaiən]n. 二价阴离子
dichloride[dai'klɔ:raid]n. 二氯化物
die[dai]n. 模头,口模
dielectric[ˌdaii'lektrik]adj. 非传导性的,诱电性的;n. 介电,绝缘体
diene['daii:n]n. 二烯(等于 diolefin)
differential[difə'renʃ(ə)l]adj. 微分的,差别的,特异的;n. 微分,差别
diffuse[di'fju:z]v. 扩散
diffusion[di'fju:ʒən]n. 扩散,漫射
difunctional[dai'fʌŋkʃənəl]adj. 双官能度的
diisocyanate[daiˌaisə'saiəneit]n. 二异氰酸酯
dilatant[dai'leitənt]n. 膨胀物 adj. 膨胀

的,因膨胀而变形的
dilatometer[ˌdilə'tɔmitə]*n.* [仪]膨胀计
dilatometry[dilə'tɔmitri]*n.* [分化][物]膨胀测定法,膨胀法
dilute[dai'lju:t]*adj.* 稀,经稀释的
dimension[di'menʃən]*n.* 尺寸,度量,方面,部分
dimensional[di'menʃənl]*adj.* 尺寸的,有尺寸的
diol['dai'ɔl]*n.* 二醇
dipole['daipəul]*n.* 偶极
dispenser[dis'pensə]*n.* 喷洒器
dispersion[dis'pə:ʃən]*n.* 分散
disposable[dis'pəuzəbl]*adj.* 用完即可丢弃的
disproportionation[disprəpə:ʃə'neiʃən]*n.* 不均,不对称,氢原子转形,歧化
dissipate['disipeit]*vt.* 分散,消散
dissolution[disə'lju:ʃən]*n.* 溶解
dissolve[di'zɔv]*vi.* 溶解
distill[dis'til]*v.* 蒸馏,提取
distillable[dis'tiləbl]*adj.* 可由蒸馏而得的
distinct[dis'tiŋkt]*adj.* 不同的,明显的
distinction[dis'tiŋkʃən]*n.* 差别,对比,区分
distinguish[dis'tiŋgwiʃ]*v.* 特性,区分
distorted[dis'tɔ:tid]*adj.* 扭歪的,受到曲解的
distortion[dis'tɔ:ʃən]*n.* 变形,畸变,挠曲
distributor[dis'tribjutə]*n.* 发行人,分散
distributor[dis'tribjutə]*n.* 配电盘
dot[dɔt]*n.* 点,圆点
dotted['dɔtid]*adj.* 点线的
downspout['daunspaut]*n.* 水落管
drape[dreip]*v.* 悬挂
draw[drɔ]*v.* 拉,拽
drawing['dʒɔ:iŋ]*n.* 拉伸
durability[ˌdjuərə'biliti]*n.* 经久,耐久力
durable['djuərəbl]*adj.* 持久的,耐用的
dye[dai]*v.* 染色
dynamic[dai'næmik]*adj.* 动力学的
ebonite['ebənait]*n.* 硬质橡胶
eccentricity[eksen'trisiti]*n.* 偏心,古怪,[数]离心率
elasticity[læsi'tisiti]*n.* 弹性,弹力
elastomer[i'læstəumə]*n.* 弹性体
electron[i'lektrɔn]*n.* 电子
electronegativity[i'lektrəuˌnegə'tiviti]*n.* 负电性
electrostatically[i'lektrəuˌstætiks]*n.* 静电学
element['elimənt]*n.* 要素,元素,成分,元件,自然环境
elevated['eliveitid]*adj.* 提高的,升高的
elimination[iˌlimi'neiʃən]*n.* 消除
elongation[ˌilɔ:ŋ'geiʃən]*n.* 伸长率
elsewhere['els'hwɛə]*adv.* 在别处,到别处
elucidate[i'lju:sideit]*vt.* 阐明
embedding[im'bediŋ]*n.* 镶铸
emboss[im'bɔs]*vt.* 压纹,压花
emerge[i'mə:dʒ]*vi.* 浮现,形成
empiricism[em'pirisizəm]*n.* 经验主义
emulsifier[i'mʌlsifaiə]*n.* 乳化剂
emulsion[i'mʌlʃən]*n.* 乳状液
encapsulation[inˌkæpsju'leiʃən]*n.* 封铸,封装
enclose[in'kləuz]*vt.* 包裹
endothermal[ɛndo'θərməl]*adj.* 吸热的
endothermic[ˌendəu'θʒ:mik]*adj.* [热]吸热的,温血的
enhance[in'ha:ns]*v.* 提高,增加
enormously[i'nɔ:məsli]*adv.* 非常地,巨大地
entangle[in'tæŋgl]*vt.* 使纠缠,卷入,使混乱
enthalpy[en'θælpi]*n.* [热]焓,[热]热函,热含量

entrench[in'trentʃ] v. 牢固树立,确定
entropy['entrəpi] n. 熵
EPDM n. 三元乙丙橡胶
epoxy[e'pɔksi] adj. 环氧的
equilibrium[ˌi:kwi'libriəm] n. 平衡,均势
equilibrium[ˌi:kwi'libriəm] n. 均衡,平静,保持平衡的能力
equivalence[i'kwivələns] n. 等价,等值
equivalent[i'kwivələnt] adj. 相当的,相等的
essentially[i'senʃəli] adv. 本质上
ester['estə] n. 酯,酯基
esterification[esˌterifi'keiʃən] n. 酯化作用
evaporate[i'væpəreit] v. 蒸发,失去水分,消失
evaporated[i'væpəreitid] adj. 浓缩的,脱水的,蒸发干燥的
evaporation[iˌvæpə'reiʃən] n. 干燥,蒸发
evolution[evə'lu:ʃən] n. 改进,进展
evolve[i'vɔlv] vt. ;vi. . 进展,进化,展开
exacting[ig'zæktiŋ] adj. 需细致小心的,(标准)严格的,难达到的
exceed[ik'si:d] vt. 超过,胜过
exclusively[ik'sklu:sivli] adv. 仅仅
exert[ig'zə:t] vt. 发挥,施加
exhaustion[ig'zɔ:stʃən] n. 消耗,用完
exothermal[ˌeksəu'θə:məl] adj. 放热的,放能的
expand[iks'pænd] n. 膨胀
expanded[iks'pændid] adj. 膨胀的
expansion[iks'pænʃən] n. 膨胀
exponential[ˌekspəu'nenʃəl] adj. 幂律的
extinguish[iks'tiŋgwiʃ] v. 熄灭
extract[iks'trækt] vt. & n. 提取,榨出,引用,提炼物,浓缩物
extreme[iks'tri:m] n. 极端,极端的;adj. 末端的
extrudate[iks'tru:deit] n. 挤出型材
extrudate[iks'tru:deit] n. 挤出物,挤出料
extrusion[eks'tru:ʒən] n. 挤出

fabric['fæbrik] n. 织物,纤维织物
fabricating['fæbrikeitiŋ] n. 二次加工
facilitate[fə'siliteit] vt. 使容易,便于
fantastic[fæn'tæstik] adj. 奇异的,幻想的
feedstock['fi:dstɔk] n. 胶料,原料
fiber['faibə] n. 纤维
fibril['faibril] n. 微纤
filament['filəmənt] n. 细丝
filter['filtə] n. 滤波器,过滤器,滤色镜;v. 过滤,渗透
flake[fleik] n. 片状
flammable['flæməbl] adj. 易燃的,可燃性的
flex[fleks] n. 挠曲
flexibility[ˌfleksə'biliti] n. 灵活性,柔韧性,弯曲性,挠性
flexible['fleksəbl] adj. 柔韧的
float[fləut] v. 浮起
fluidize['fluidaiz] vt. 流化
fluorine['fluəri:n] n. 氟
fluorocarbon[ˌflu:ərəu'ka:bən] n. 碳氟化合物
foam[fəum] v. 发泡; n. 泡沫
folding['fəuldiŋ] n. 折叠
formability[fɔ:mə'biliti] n. 可成型性
formaldehyde[fɔ:'mældiˌhaid] n. 甲醛,蚁醛
formica[fɔ:maikə] n. 胶木
formulation[fɔmju:'leiʃən] n. 配方
foundry['faundri] n. 铸造厂
fraction['frækʃən] n. 分散,小部分,片段,分数
fractionate['frækʃəneit] v. 分级,分馏
fracture['fræktʃə] n. 破裂,断裂
fragment['frægmənt] n. 碎片,碎屑
frame[freim] n. 模架
frequency['fri:kwənsi] n. 频率
functionality[ˌfʌŋkəʃə'næliti] n. 官能度
fundamental[ˌfʌndə'mentl] n. 基本原理;adj. 基本的,根本的

fusing['fju:ziŋ]*n.* 熔融,塑化
fusion['fju:ʒən]*n.* 熔结,熔融
gasket['gæskit]*n.* 衬垫,垫圈,垫片
gate[geit]*n.* 浇口
gauche[gəuʃ](conformation) 邻位交叉(构象),扭曲(构象)
gauge[gedʒ]*n.* 标准尺,规格,量规,量表;*v.* 测量
gear[giə]*n.* 齿轮
generalization[ˌdʒenrələ'zeiʃən]*n.* 概括,普遍化,一般化
geometric[dʒi'ɔmətri]*adj.* 几何的
geometrically[ˌdʒiə'metrikəli]*adv.* 几何学上
gigapascal[gigə 'pæsk(ə)l]*n.* 吉帕斯卡(物理单位)
glassy['glɑ:si]*adj.* 像玻璃的;光亮透明的,呆滞的
globe[gləub]*n.* 球体 球形物
glycol['glaikəl]*n.* 乙二醇
gradually['grædjuəli]*adv.* 逐渐地
granular['grænjulə]*adj.* 粒状的,颗粒状的
granulating[ˌgænju'leitiŋ]*n.* 造粒
granule['grænju:l]*n.* 颗粒
grill[gril]*n.* 烤架,烤肉;*v.* 烧,烤
groove[gru:v]*n.* 凹槽
gum[gʌm]*n.* 胶
gusset['gʌsit]*n.* 边折,衣袖
gutter['gʌtə]*n.* 水槽
gyration[dʒaɪ'reʃən]*n.* 旋转,[力]回转
gyroplane['dʒaiərəplein]*n.* 旋翼机
hallmark['hɔ:l'ma:k]*n.* 标记
hardener['ha:dnə]*n.* 固化剂
hardness['ha:dnis]*n.* 硬度
hardware['ha:dwɛə]*n.* 金属构件
helix['hilIks]*n.* 螺旋
helmet['helmit]*n.* 头盔
hide[haid]*n.* 兽皮,皮革
hinge[hindʒ]*n.* 铰链
hobby['hɔbi]*n.* 切压
hollow['hɔləu]*adj.* 中空的
homogeneous[ˌhɔməu'dʒi:njəs]*adj.* 均匀的,均一的
homogenization[hə ˌmɔdʒənai 'zeiʃən]*n.* 均匀化
homopolymer[ˌhəumə'pɔlimə]*n.* 均聚物
hoof[hu:f]*n.* (兽的)蹄,马蹄
hopper['hɔpə]*n.* 料斗
hose[həuz]*n.* 水管,橡皮软管,胶管;*vt.* 用软管浇(冲洗)
houseware['hauswɛə]*n.* 家庭用具
hydraulic[hai'drə:lik]*adj.* 液压的
hydraulically[hai'drə:likli]*adv.* 液压地
hydrocarbon['haidrəu'ka:bən]*n.* 烃,碳氢化合物
hydrogen['haidrədʒən]*n.* 氢,氢气
hydrolysis[hai 'drɔlisis]*n.* 水解,水解作用
hydrophilic[ˌhaidrəu'filik]*adj.* 亲水的,吸水的
hydroxyl[hai'drɔksil]*n.* 羟(基)
hypothesis[hai'pɔθisis]*n.* 假设
hysteresis[ˌhistə'ri:sis]*n.* 迟滞现象,滞后作用,磁滞现象
identical[ai 'dentikəl]*adj.* 相同的,一致的
identify[ai'dentifai]*vt.* 鉴别
immerse[i'mə:s]*vt.* 浸,陷入
immiscible[i'misib(ə)l]*adj.* 不融和的,不能混合的
immobilize[i'məubilaiz]*vt.* 不能活动
impregnate['impregneit]*n.* 浸渍树脂;*v.* 浸渍
impurity[im'pjuəriti]*n.* 杂质
inactive[in'æktiv]*adj.* 不活泼的
incineration[inˌsinə'reiʃən]*n.* 焚化,烧尽
incorporate[in'kɔ:pəreit]*vi.* 合并,合并的;*adj.* 具体化的
incorporated[in'kɔ:pəreitid]*adj.* 组成公

司的,合成一体的,合并的
indicate['indikeit] vt. 指示,表明
indispensable[indis'pensəbl] adj. 不可缺少的
inert[i'nə:t] adj. 惰性的
inertia[i'nə:ʃiə] n. 惯性,惯量
infusible[in'fju:zəbl] adj. 不溶解的
ingredient[in'gri:djənt] n. (混合物的)组成部分,配料
inherent[in'hiərənt] adj. 固有的
initial[i'niʃəl] adj. 初步,最初
initiate[i'niʃieit] vt. 引发,开始,发动,传授;v. 开始,发起
initiation[iˌniʃi'eiʃən] n. 开始,引发
injection[in'dʒekʃən] n. 注射
innovation[inəu'veiʃən] n. 创新,改革
innovative['inəuveitiv] adj. 新发明的,新引进的,革新的
inroad['inrəud] n. 改进
insignificant[ˌinsig'nifikənt] adj. 无关紧要的,可忽略的
inspection[in'spekʃən] n. 检查
instantaneously[ˌinstən'teiniəsli] adv. 霎时,立即
insulation[ˌinsju'leiʃən] n. 绝缘
intact[in'tækt] adj. 完整的,整体的
integral['intigrəl] adj. 完整的,整体的
integration[ˌinti'greiʃən] n. 集成
intensive[in'tensiv] adj. 集中的,强化的,深入的
intentionally[in'tenʃənli] adv. 有意地,故意地
interaction[ˌintər'ækʃən] n. 相互作用,相互影响
interchain[ˌintə'tʃein] adj. 链间的
interdependent[ˌintə:di'pendənt] adj. 相互关联的
interfacial[ˌintə(:)'feiʃəl] adj. 界面的
interlamellar[ˌintə(:)lə'melə] adj. 层间的
intermesh[ˌintə(:)'meʃ] vt. 使互相结合,使互相啮合
intermolecular[ˌintə(:)mə'lekjulə] adj. 分子间的,存在(或作用)于分子间的
interstice[in'tə:stis] n. 间隙,空隙
intrigue[in'tri:g] v. 激起……的兴趣
invariably[in'vɛəriəbli] adv. 不变地,总是
invariant[in'vɛəriənt] adj. 不变的
invoke[in'vəuk] vt. 援引
ionic[ai'ɔnik] adj. 离子的
ionize['aiənaiz] vt. 使离子化,vi. 电离
ionomer[ˌaiə'nəmə] n. 含离子键的聚合物
irreversible[ˌiri'və:səbl] adj. 不可逆的,不能撤回的,不能取消的
irreversibly[ˌiri'və:səbli] adv. 不可逆地,单向地
isobutylene[ˌɑisəu'bju:tili:n] n. [化]异丁烯
isochronal[ai'sɔkrənəl] adj. 等时线的
isocyanate[ˌaisəu'saiəneit] n. 异氰酸酯
isoprene[ai'səupri:] n. 异戊二烯
isotactic[ˌaisəu'tæktik] adj. 全同立构的,等规立构的
isotactic[ˌaisəu'tæktik] adj. 等规的,全同立构的,等规立构的
ivory[ˌɑivəri] n. 象牙
jaw[dʒɔ:] n. 夹头
jet[dʒet] n. 喷流,射流
jewelry['dʒuəlri] n. 宝石 珍宝
jig[dʒig] n. 夹具
judicious[dʒu:Ⅱ'diʃəs] adj. 明智的
ketchup['ketʃəp] n. 番茄酱
lamella[lə'melə] n. 薄板,薄片,薄层
lamellae[læ'meli:] n. 晶片
laminar['læminə] adj. 薄片状的,层状的
laminate['læmineit] v. 切成薄板(片);n. 层压板,层压材料
laminating['læmineitiŋ] n. 层压,层合

landfill['lændfil]n. 垃圾,垃圾掩埋法
latex['leiteks]n. 胶乳,乳胶
lattice['lætis]n. 格子
lead[li:d]n. (螺杆)导程,导线,铅;vt. & vi. 引导,致使,导致
lengthwise['leŋθwaiz]adj. & adv. 纵向的(地)
linear['liniə]adj. 线型的
location[ləu'keitʃən]n. 定位,地点,位置
locus['ləukəs]n. 轨迹
logarithmic[ˌlɔgə'riθmik]adj. 对数的
lubricant['lju:brikənt]n. 润滑剂
lubricate['lju:brikeit]v. 上油,使润滑
lubrication[ˌlu:bri'keiʃən]n. 润滑油
lyotropic[lai'ɔtrɔpik]adj. 溶致的,易溶的
Macbeth[mək'bəθ]n. 麦克佩斯
machinability[mə'ʃi:nəbiliti]n. 机械加工性
magnesium[mæg'ni:zjən]n. 镁
magnitude['mægnitju:d]n. 大小,尺寸,巨大,重要性,程度
makeup['meikʌp]n. 组成(接通,补给,修理)
mandrel['mændrəl]n. 芯棒,芯模
manifest['mænifest]vt. 表明,表现,证明
manipulation[məˌnipju'leiʃən]vt. 处理,操作,控制
marble['ma:bl]n. 大理石
markedly['ma:kidli]adv. 明显地
masonry['meisənri]n. 建筑
mastication[ˌmæsti'keiʃən]n. 塑炼,捏炼
matrix['mætriks]n. 基体
maximize['mæksimaiz]vt. 最大化
MBTS n. 促进剂 MBTS
mechanical[mi'kænikl]adj. 机械的,力学的
melamine['meləmi(:)n]n. 三聚氰(酰)胺,密胺
mer['mə:]n. 基体
mesogenic[mesəu'dʒenik]adj. 液晶的
mesomorphic[ˌmesəuˌmɔ:fik]adj. 具有中间相的;
mesophase['mesəufeiz]n. 中间相,液晶相
metal['metl]n. 金属;adj. 金属制的
metering['mi:təriŋ]n. 测量(法),计量,测定
methacrylate[me'θækrəleit]n. 甲基丙烯酸酯
methacrylic[me'θəkrilik]adj. 甲基丙烯类的
methyl['meθil]n. 甲基
micelle[mi'sel]n. 胶束,胶囊,微胞,微团,胶态离子
milestone['mailstəun]n. 里程碑
milky['milki]adj. 乳状的
mill[mil]n. 研磨机
minimize['minimaiz]vt. & vi. 最小化
miscible['misibl]adj. 易混的
misleading[mis'li:diŋ]adj. 使人误解的
mixer['miksə]n. 混炼机,混合机
modacrylic[ˌmɔdə'krilik]n. 改性聚丙烯腈
model['mɔdl]n. 模塑
moderately['nɔdəritli]adv. 适中地
modification[ˌmɔdifi'keiʃən]n. 修饰,改性
modifying['mɔdifaiiŋ]n. 改性
moisture['mɔisʃə]n. 水分
molecule['mɔliku:l]n. 分子
molten['məultən]adj. 熔化的,炽热的,铸造的
monarch[mɔk'nək]n. 主要,统治,君主
monofilament['mɔnəu'filəmənt]n. 单(根长)丝,单纤(维)丝
monomer['mɔnəmə]n. 单体
morphology[mɔ:'fɔlədʒi]n. 形态,形态学
motion['məuʃən]n. 动力,运动
mount[maunt]vt. 固定,安装

MRX *n.* 矿质橡胶
multifilament[ˌmʌlti'filəmənt] *n.* 复丝,多纤(维)丝
mutual['mjuːtjuəl] *adj.* 相互的,共同的
neglect[ni'glekt] *vt.* 疏忽,忽视;*n.* 疏忽,忽略
nematic[ni'mætik] *adj.* (液晶)[晶体]向列的
neoprene['niəupriːn] *n.* 氯丁胶
nerve[nəːv] *n.* 回缩性
network['netwəːk] *n.* 网络
neutral['njuːtrəl] *adj.* 中性的,无确定性质的,(颜色等)不确定的
nip[nip] *n.* 夹缝
nitrate['naitreit] *n.* 硝酸盐
nitrile['naitriːl] *n.* 丙烯腈,腈类
nitrocellulose[naitrəu'seljuləus] *n.* 硝化纤维,硝酸纤维素
nitrogen['naitrdʒən] *n.* 氮
nomenclature[nəu'menklətʃə] *n.* 命名法,命名原则
nonbreakable['nɔnbreikəbl] *adj.* 不易碎的
nonperishable[nɔn'periʃəbl] *adj.* 不易腐烂的
nook[nuk] *n.* 隐蔽处,死角
notably['nəutəbli] *adv.* 尤其是,值得注意地
notation[nəu'teiʃən] *n.* 记号法,表示法
noticeably['nəutisəbli] *adv.* 显著地,明显地
novelty['nɔvəlti] *n.* 新产品
nozzle['nɔzl] *n.* 喷嘴
NR *n.* 天然橡胶
nylon['nailən] *n.* 尼龙
object[əb'dʒekt] *v.* 反对
oligomer[ə'ligəmə] *n.* 低聚物,低聚体
omission[əu'miʃən] *n.* 省略,遗漏,疏忽
opaque[əu'peik] *adj.* 不透明的
opening['əupniŋ] *n.* 加料口
optical['ɔptikəl] *adj.* 眼睛的,视觉的,光学的
optronics[ɔp'trɔniks] *n.* 光电子学,光导发光学,光电产品
orient['ɔːriənt] *v.* 取向
orientation['ɔːrien'teiʃən] *n.* 取向;*adj.* 方向,目标
originator[ə'ridʒəneitə] *n.* 起源,起因,来源
oscillation[ˌɑːsi'leiʃn] *n.* 振动;波动;动摇;<物>振荡
outline['əutlain] *n.* 大纲;*vt.* 概要,描述要点
overlapping[ˌəuvə'læpiŋ] *n.* 复叠,重叠
overrun[ˌəuvə'rʌn] *n.* 超出限度;*vt.* & *vi.* 泛滥
overwhelming[ˌəuvə'hwelmiŋ] *adj.* 优势的
ozone['əuzəun] *n.* 臭氧
paint[peint] *n.* 颜料,油漆
palletizing['pelitaiziŋ] *n.* 微粒化
paraffin['pærəfin] *n.* 石蜡 烷烃
parameter[pə'ræmitə] *n.* 参数,参量
parison['pærisən] *n.* 型坯
parylene['paːriliːn] *n.* 聚对二甲苯
paste[peist] *n.* 糊(塑料)
pedal['pedl] *n.* 踏板
pellet['pelit] *n.* 粒料,切粒
pelletize['pelitaiz] *vt.* 制成颗粒,造粒
pendulum[ˌpendjuləm] *n.* 钟摆,摇锤
penetrate['penitreit] *vt.*;*vi.* 穿透,渗透
penultimate[pi'nʌltimit] *adj.* 倒数第二个音节的
perduren[pəː'djuːrən] *n.* 硫化橡胶
periodicity[piəriə'disiti] *n.* 周期性
peroxide[per'ɔksaid] *n.* 过氧化物
perpendicular[ˌpəːpən'dikjulə] *adj.* 直立的,垂直
petrochemical[ˌpetrəu'kemikəl] *n.* 石化产品;*adj.* 石化的
petroleum[pi'trəuliəm] *n.* 石油

phase[feiz]n. 阶段
phenol['fi:nɔl]n. 苯酚
phenolic[fi'nɔlik]n. 酚醛塑料,酚醛树脂,酚类(的)
phenoxy['finɔksi]n. 苯氧基
phenyl['fenil]n. 苯基
phosphate['fɔsfeit]n. [无化]亚磷酸盐
photodiode[fəutəu'daiəud]n. [电子]光电二极管
phthalate[θæleit]n. 酞酸盐(酯)
picnic['piknik]n. 野餐
pigment['pigmənt]n. 颜料
piston['pistən]n. 活塞
pitch[pitʃ]n. 螺距,程度,斜度;vt. & vi. 定位于,用沥青涂,投掷,倾斜
plaster['pla:stə]n. 石膏;vt. 涂以灰泥,敷以膏药
plastic['plæstik]n. 塑料; adj. 塑性的
plasticize['plæstisaiz]vt. 增塑(塑炼),塑化
plasticizer['plæstisaizə]n. 增塑剂
pliable['plaiəbl]adj. 易弯的,柔韧的,柔顺的
pliobond[pli'ɔbɔnd]n. 合成树脂结合剂
plot[plɔt]vt. 策划,绘图
plunger['plʌndʒə]n. 阳模,柱塞
plywood['plaiwud]n. 夹板,合板,胶合板
pneumatic[nju(:)'mætik]n. &adj. 气胎,气动的
polar['pɔlə]adj. 极性的
polarizability['pəulə,raizə'biləti]n. [电子]极化性,[电磁]极化度
polarization[,pəulərai'zeiʃən]n. 极化,偏振,两极分化
polished['pɔliʃt]adj. 擦亮的,抛光的
polyacrylonitrile[,pɔli,ækriləu'naitril]n. 聚丙烯腈
polyallomer['pɔli'ælɔmə]n. 同质异晶聚合物
polyamide['pɔli'æmaid]n. 聚酰胺
polybutadiene['pɔli,bju:tə'daIi:n]n. 聚丁二烯
polycarbonate[,pɔli'ka:bənit]n. 聚碳酸酯
polydisperse[,pɔlidis'pə:s]adj. 多分散性的
polyester[,pɔli'estə]n. 聚酯
polyethylene[,pɔli'eθili:n]n. 聚乙烯
polyisobutylene[,pɔliaisəu'bju:tilin]n. 聚异丁烯
polyisoprene[,pɔli'aisəupri:n]n. 聚异戊二烯
polylactone[,pɔli'læktəun]n. 聚内酯
polymer[,pɔlimə]n. 聚合物
polymerization[,pɔliməraі'zeiʃən]n. 聚合,聚合反应
polymerize[,pɔliməraiz]vi. (使)聚合
polymorphs[,pɔlimɔ:fs]n. [晶体]多形体
polyolefin[,pɔli'əuləfin]n. 聚烯烃
polyoxymethylene[,pɔli,ɔksi'meθili:n]n. 聚甲醛
polypropylene[,pɔli'prɔpilin]n. 聚丙烯
polystyrene[,pɔli'staiəri:n]n. 聚苯乙烯
polysulfone[,pɔli'sʌlfəun]n. 聚砜
polytetrafluoroethylene[,pɔli,tetrə,fluərə'eθili:n]n. 聚四氟乙烯
polyurethane[,pɔli'juəriθein]n. 聚氨酯,聚氨基甲酸酯
pontoon[pɔn'tu:n]n. 浮桥(船)
porous['pɔ:rəs]adj. 可渗透的,多孔的,疏松的
port[pɔ:t]n. 端口,港口
pot[pɔt]n. 罐
potassium[pə'tæsiəm]n. 钾
precaution[pri'kɔ:ʃən]n. 预防措施
precipitate[pri'sipiteit]vi. 沉淀,析出
precision[pri'siʒən]n. 精密度
predetermine['pri:di'tə:min]vt. 预定,预先确定
predominate[pri'dɔmineit]vt. 起主导作

用,居支配地位
prerequisite['pri:'rekwizit] *n.* 先决条件; *adj.* 首先具备的
presumably[pri'zju:məbəli] *adv.* 推测,大概
prewarm['pri:'wɔ:m] *v.* 预热
proceed[prɔ'sid] *vi.* 继续进行
progressively[prə'gresivli] *adv.* 逐渐地
propagate['prɔpəgeit] *v.* 繁殖,传播,宣传
propagation[ˌprɔpə'geiʃən] *n.* 增长,繁殖
propeller[prə'pelə] *n.* 螺旋桨
property['prɔpəti] *n.* 性能,性质
proportional[prə'pɔ:ʃɔnl] *n.* 比例的,成比例的
propylene['prəupili:n] *n.* 丙烯
protein['prəuti:n] *n.* 蛋白质
prototype['prəutətaip] *n.* 原型
protruding[prə:'tru:diŋ] *adj.* 凸,突出
pseudoplastic[psju:dəu'plæstik] *adj.* 假塑性的
puddle['pʌdl] *n.* 胶泥;*vt.* 搅拌
pump[pʌmp] *n.* 抽水机,泵;*vi.* 抽水,打气
purification[ˌpjuərifi'keiʃən] *n.* 净化,提纯
putty['pʌti] *n.* 油灰,氧化锡;*v.* 用油灰填塞
pyramidal[pi'ræmidl] *adj.* 锥体的
pyrolysis[pai'rɔlisis] *n.* 热解作用,高温分解
pyrometer[pai'rɔmitə] *n.* 高温计
qualitatively['kwɔlitətiv] *adv.* 定性
quantitatively['kwɔntitətivli] *adv.* 定量地
quantity['kwɔntəti] *n.* 数量
quartz[kwɔ:ts] *n.* 石英
radiant['reidiənt] *adj.* 放热的,发光的,辐射的
random['rændəm] *adj.* 无规的
rationalize[ræʃənlaiz] *v.* 合理地说明
reaction[ri(:)'ækʃən] *n.* 反应
recipe['resipi] *n.* 配方,处方
reentry[ri:'entri] *n.* 折返
refinery[ri'fainəri] *n.* 精炼厂
regain[ri'gein] *vt.* 恢复,重回
regrind[ri'graund] *v.* 再粉碎,再研磨
regrinding[ri'graindiŋ] *v.* &*n.* 再粉碎
reject['ri:dʒekt] *n.* 废料
remold['ri:'məuld] *vt.* 改造,重铸
reservoir['rezəvwa:] *n.* 堆积,储藏
residues[ri'zidjuəm] *n.* 残余,剩余物,残数,滤渣,渣滓
residuum[ri'zidjuəm] *n.* 剩余,残滓
restriction[ris'trikʃən] *n.* 限制,限定
reversible[ri'və:səbl] *adj.* 可逆的
reversibly[ri'və:səbli] *adv.* 可逆地
rheological[ˌriə'lɑdʒikəl] *adj.* 流变学的,液流学的
rheological[ˌriə'lɔdʒikl] *adj.* 流变学的
rheopectic[ˌri:ə'pektik] *adj.* 振凝的,抗流变的
rigid['ridʒid] *adj.* 僵硬的,刻板的,严格的
rigidity[ri'dʒiditi] *n.* 刚性,刚度
rigorous['rig(ə)rəs] *adj.* 严格的,严厉的,严密的
rigorous['rigərəs] *adj.* 严密的,缜密的
rivet['rivit] *n.* 铆钉,*vt.* 铆接
robbin['rɔbin] *n.* 筒,管
rod[rɔd] *n.* 杆,棒
roller['rəulə] *n.* 辊子,滚轴
rotate[rəu'teit] *vt.* ;*vi.* (使)旋转
rotational[rəu'teiʃənl] *n.* 滚塑
rubber['rʌbə] *n.* 橡胶
rubbery['rʌbəri] *adj.* 强韧的
rugged['rʌgid] *adj.* 高低不平的,崎岖的,粗糙的,有皱纹的
runner['rʌnə] *n.* 流道
sap[sæp] *n.* 树汁
sawing['sɔ:iŋ] *n.* 锯,锯切,锯开
SBR *n.* 丁苯橡胶

scale[skeil]*n.* 规模
scarce[skεəs]*adj.* 缺乏的,不足的
schematically[ski'mætikli]*adv.* 示意地,大略地
scission['siʒən]*n.* 剪断,分隔
scope[skəup]*n.* 范围,机会,广度,眼界
scorching['skɔ:tʃiŋ]*adj.* 灼热的,激烈的
scrap[skræp]*n.* 残余物;*vt.* 扔弃
scratch[skrætʃ]*n.* 刮伤
screw[skru:]*n.* 螺杆,螺丝钉;*vt.* 调节,旋;*vi.* 转动,旋,拧
section['sekʃən]*n.* 型材
segmental[seg'mentəl]*adj.* 部分的
sensitive['sensitiv]*adj.* 敏感的
shake[ʃeik]*n.* 摇动,震动
shank[ʃæŋk]*n.* 根部,后部,胫,腿骨
sheath[ʃi:θ]*n.* 鞘,套
shock[ʃɔk]*n.* 冲击,振动
shot[ʃɔt]*n.* 注射,注射量
siding['saidiŋ]*n.* 板壁
silicone['silikəun]*n.* 硅,硅有机树脂
similarity[ˌsimi'læriti]*n.* 相似,类似,相似性,像
simplicity[sim'plisiti]*n.* 简单,简单性,单一性
simultaneously[siməl'teiniəsli]*adv.* 同时发生,一起,同时,同时存在
sinter['sintə]*n.* & *v.* 烧结
slide[slaid]*vi.* 滑,跌落
slippage['slipidʒ]*n.* 滑移,滑动
slit[slit]*vt.* 切开
slope[sləup]*n.* 斜率
slurries['slə:ri]*n.* 泥浆
smectic['smektik]*adj.* 近晶的;净化的
snap[snæp]*vt.* 折断,咬断,按扣
soak[səuk]*vi.* 浸润
sodium['səudjəm]*n.* 钠
soffit['sɔfit]*n.* 下端背面,拱腹
soft[sɔft]*adj.* 柔软的
solidify[sə'lidifai]*vt.* 使凝固,硬化
solubility[ˌsɔlju'biliti]*n.* 溶解性
soluble['sɔljubl]*adj.* 溶解的,可溶解的
solute['sɔlju:t]*n.* 溶质
solution[sə'lu:ʃən]*n.* 溶液
solvent['sɔlvənt]*n.* 溶剂
somewhat['sʌmwɔt]*adv.* 稍微,有点
spaghetti[spə'geti]*n.* 意大利面条
spark[spa:k]*n.* &*vi.* &*vt.* 火花
specific[spi'sifik]*adj.* 特定的
specification[ˌspesifi'keiʃən]*n.* 技术要求
specifies['spi:ʃi:z]*n.* 种类,类型
specimen['spesimin]*n.* 样品,样本,标本
spectrum['spektrəm]*n.* 光,光谱 ,型谱,频谱
sphere[sfiə]*n.* 球,球体,范围,领域,方面,圈子,半球
spherulite['sferjulait]*n.* 球晶,[地质]球粒
spike[spaik]*n.* 峰,峰形
spin[spin]*v.* 纺
spinneret[ˌspinə'ret]*n.* 喷丝头,纺丝头
spinnerette[ˌspinə'ret]*n.* 纺丝头,喷丝头
spinning['spiniŋ]*n.* 纺织,纺纱
splash[splæʃ]*n.* 溅,飞溅,斑点;*v.* 溅,泼,溅湿
split[split]*n.* 裂口,薄片
sponge[spʌndʒ]*n.* 海绵
spring[spriŋ]*n.* 弹簧
sprue[spru:]*n.* 注道
squeeze[skwi:z]*v.* ;*n.* 挤压
stabilizer['steibilaizə]*n.* [助剂]稳定剂,稳定器
stable['steibl]*adj.* 稳定的
stactic[e'tæktIk]*adj.* 不规则的,无规立构的
staggered position['stægəd]*adj.* 错列开的位置
stainable['steinəbl]*adj.* 可着色的
stamping['stæmpiŋ]*v.* 冲压;*n.* 压膜
staple['steipl]*n.* 纤维产品

starch[sta:tʃ] n. 淀粉
stem[stem] vi. 起源,发生
stepwise['stepwaiz] ad. 逐步的
stereospecific[ˌstiəriəuspə'sifik] adj. 有规立构定向的
stereostructure['stɛriostɛrio] n. 立体结构
sterilizable[ˌsterə'lizəbl] adj. 杀菌的
sticky['stiki] adj. 黏的,黏性的
stiff[stif] adj. 刚性的
stimuli['stimjulai] n. stimulus 的复数,刺激因素,促进因素
stock[stɔk] n. 毛坯,坯料
stoichiometry[ˌstɔiki'ɔmitri] n. 化学计算(法),化学计量学
strain[strein] n. 应变
strand[strænd] n. 线料
stretch[stretʃ] vi. & vt. 拉伸,伸展
stripper['stripə] n. 脱模板
styrene['staiəri:n] n. 苯乙烯
subclass['sʌbklɑ:s] n. 子类,子集
subdivision[ˌsʌbdi'viʒən] n. 细分,分段
submerge[səb'mə:dʒ] vt. , & vi. (使)潜入水中,淹没,完全掩盖,遮掩
subscript[səb'skrip] n. 下标
substantial[səb'stænʃəl] adj. 很多的,物质的
substantiate[səb'stænʃieit] vt. 证明,证实
sulfate['sʌlfeit] n. 硫酸盐;vt. 以硫酸或硫酸盐处理,使变为硫酸盐
sulfur['sʌlfə] n. 硫
superheating[ˌsju:pə'hi:tiŋ] n. [热]过热,市场狂热;v. 过度加热
superposition[ˌsju:pəpə'ziʃən] n. [数]叠加,重合
swell[swel] vi. 溶胀
symmetrical[si'metrikəl] adj. 对称的,匀称的
syndiotactic [ˌsindaɪu'tæktɪk] adj. 间同(立构)的,间规(立构)的
synthesis['sinθisis] n. 合成,综合
synthetic[sin'θetik] adj. 合成的,人造的
syrup['sirəp] n. 树脂浆
systematic[ˌsistə'mætik] adj. 系统的,有体系的
tack[tæk] n. 黏性
tacticity[tæk'tisəti] n. 立构规整度
takeup['teik'ʌp] n. 卷绕
tan[tæn] n. 棕褐色
taper['teipə] n. 锥形,锥度;v. 逐渐变细,逐渐减少
tapering ['teipəriŋ] adj. 尖端细的,锥形的
tar[ta:] n. 焦油
tear[tɛə] v. & n. 撕裂
technologist[tek'nɔlɔdʒist] n. 工艺师
temporary ['tempərəri] adj. 暂时的,临时的
tendency['tendənsi] n. 趋势
tensile['tensail] adj. 可拉伸的
terminal['tə:minl] n. 端基
termination[tə:mi'neiʃən] n. 终止,端基
terminology[ˌtə:mi'nɔlədʒi] n. 专业术语
terpolymer[tə:'pɔlimə] n. 三元共聚物
tetrafluoroethylene[ˌtetrəˌfluərə'eθili:n] n. 四氟乙烯
textile['tekstail] n. 纺织材料
texture['tekstʃə] n. (织物)组织,构造;vt. 使具有某种结构(组织)
theorem['θiərəm] n. [数]定理
thermobalance[ˌθə:məu'bæləns] n. [分化]热天平
thermodynamics[ˌθə:məudai'næmiks] n. 热力学
thermoform['θə:məufɔm] n. 热成型;vt. 使加热成型,给……用热力塑型
thermogravimetry[ˌθɜ:məugrə'vi1mi1tril] n. [分化]热重量分析法
thermometry[θə'mɔmitri] n. 温度测量,温度测定法,[物]计温学
thermopile['θɜ:mə(u)pail] n. [热]热电

电偶,[电]温差电偶
thermoplastics[ˌθə:mə'plæstiks] n. 热塑性塑料;adj. 热塑性的
thermosets['θə:məsets] n. 热固性材料
thickness['θiknis] n. 厚度
thioester[ˌθaiəu'estə] n. 硫酯,[有化]硫代酸酯
thixotropic[ˌθiksə'trɔpik] adj. 触变性,摇溶性
three-dimensional[θri:di'menʃənəl] adj. 三维的,立体的
tolerance['tɔlərəns] n. 公差
toluene['tɔljui:n] n. 甲苯
torsion['tɔ:ʃ(ə)n] n. 扭转,扭曲,转矩,[力]扭力
tough[tʌf] adj. 韧性的,刚性的
toughness['tʌfnis] n. 韧性
transfer['trænsfə:] n. 压铸,传递
translucent[trænz'lju:sənt] adj. 半透明的,透明的
transparency[træns'pεərənsi] n. 透明性
tray n. 托盘
tremendous[tri'mendəs] adj. 巨大的,惊人的
trigger['trIgə] n. 引发其他事件的一件事;vt 引发,触发;
trihalides[ˌtrai'hælaid] n. 三卤化合物
trim[trim] adj. 整齐的,整洁的;vt. 整理,修整 ,装饰
trimmer['trimə] n. 切边机
trough['trɔ:f] n. 槽,水槽,饲料槽,木钵
tubing['tju:biŋ] n. 装管,配管,管道系统
turntable['tə:nteibl] n. 转盘
turpentine['tə:pəntain] n. 松节油
twist[twist] vi. 蜷曲
ultimately['ʌltimətli] adv. 最后,最终
undedicated['ʌn'dedikeit] adj. 非专用的
underlie[ˌʌndə'lai] vt. 位于……之下,成为……的基础
unfold[ʌn'fəuld] vt. & vi. 展开,打开
uniaxial['ju:ni'æksiəl] adj. 单轴的,单轴晶体
unique[ju:'ni:k] adj. 独一无二的,特有的
unoriented[ʌn'ɔ:rientid] adj. 未取向的
unsaturated['ʌn'sætʃəreitid] adj. 没有饱和的,不饱和的
untangle[ʌn'tæŋgl] v. 解缠结
upholstery[ʌp'həulstəri] n. 室内装饰
utensil[ju:'tensl] n. 用具
utilitarian[ˌju:tili'tεəriən] adj. 有效用的,实用的
vacuum['vækjuəm] n. 真空,空间;vt. 用真空吸尘器清扫(某物)
valence['veiləns] n. 化合价
vaporization[ˌveipərai'zeiʃən] n. 汽化
vector['vektə] n. 矢量,带菌者,航线
velocity[vi'lɔsiti] n. 速度,速率
veneer[və'niə] n. 板坯
versatile['və:sətail] adj. (指工具、机器等)多用途的,多功能的
versatility[və:sə'tiliti] n. 通用性
vertical['və:tikəl] adj. 立式的,竖式的
vigorous['vigərəs] adj. 有力的,精力充沛的
vigorously['vigərəsli] adv. 充满活力地
vinyl['vainil] n. 乙烯基
viscoelastic[ˌviskəui'læstik] adj. 黏弹性的,黏弹性,黏滞弹性的
viscometer[vis'kɔmitə] n. 黏度计
viscometry[vis'kɔmitri] n. 黏度测定法
viscosity[vis'kɔseti] n. 黏度,黏性
void['vɔid] n. 空隙,间隙
volatile['vɔlətail] adj. [化学]挥发性的,不稳定的;n. 挥发物
volatiles['vɔlətail] n. 挥发组分,挥发物,挥发性物质
volume['vɔlju:m] n. 体积
vulcanization[ˌvʌlkənai'zeiʃən] n. (橡胶的)硫化

water-cooled['wɔ:təku:ld] *adj.* 用水冷却的
wear[wɛə] *v.* & *n.* 磨损
weave[wi:v] *v.* 织,编织
welding['weldiŋ] *n.* 焊接法,定位焊接
wetting['wetiŋ] *n.* ;*v.* 润湿
whiting['hwaitiŋ] *n.* 白粉,铅粉
windup['waindʌp] *n.* 收卷,卷起
witch[witʃ] *n.* 女巫
wriggle['rigl] *vi.* 蠕动,蜿蜒行进,扭动
wriggling['rigliŋ] *adj.* 蠕动
xanthate['zænθeit] *n.* 黄酸盐,黄原酸盐,黄原酸酯
yarn[ja:n] *n.* 纱,纱线

Phrases and Expressions
常用词和习惯用语索引

2-chloro-1,3-butadiene　2-氯-1,3-丁二烯
a fraction of　……的几分之一
abrasion resistance　耐磨性
abrasive grains　砂磨粒
account for　占,说明
acrylic sheet　丙烯酸酯类片材
activation energy　活化能
activation entropy　活化熵
active initiator　活性引发剂
addition polymerization　加成聚合反应
aim for...　目标是
ambient workshop temperature　室温
ammonium persulfate　过(二)硫酸铵
amorphous phase　非晶相
amorphous polymers　非晶态聚合物
angular velocity　角速度
appropriate solvent　适当溶剂(良溶剂)
aqueous phase　液相
artificial fibers　人造纤维
as a function of time　作为时间的函数
as a whole　整个
associate with　与……相关,联合
atactic polymer　无规聚合物,不规则排列聚合物
atactic polypropylene　无规立构聚丙烯
auxiliary stabilizer　辅助稳定剂
average molecular weight　平均分子量
aviation component　航空构件
"balled-up" configuration　球状构型
back-up plate　支承板
back-up　返回
balanced runner　平衡式流道
banbury mixer　密炼机
bare wire　裸线
batch feeding　投料,配合料加料
be abbreviated by　简化为,缩短为
be broken off　脱落,折断
be caused by　起因于
be concerned with　包括,与……有关
be derived from　来源于
be divided into　被分成
be divided into　分为……,归结为……
be ejected from　从……中取出
be fastened to　固定在……上
be loaded with　装(载)着……
be rated according to...　以……定等级或标定
be recognized by　由……来识别
benzoyl peroxide　过氧化苯甲酰
bicomponent fiber　二组分纤维
Bingham plastic　宾汉塑性
blow molding　吹塑模塑
blowing agent　发泡剂
blown tube　吹胀膜管,管膜
boils down to　得出……的结论,归结为……
boron trifluoride　三氟化硼
bowling ball　保龄球
bragg reflections　布拉格反射
breaker plate　多孔板
brought out　引出,生产,阐述
build up　组成,形成,聚集
bulk polymerization　本体聚合
butyl rubber　丁基橡胶
by analogy　用类比的方法,同样
cable coating　电缆涂层
cable insulation　电缆绝缘层,电缆绝缘料
calendering roll　压延辊
cast aluminum　生铝,铸铝

casting molding　铸塑模塑，铸塑成型
catalyst complex　催化剂混合物
cavity block　凹模
cell structure　泡孔结构
cellular plastics　泡沫塑料
cellular structure　微孔结构
cellulose acetate　醋酸纤维素
cellulose nitrate　硝酸纤维素
cellulose xanthate　黄酸纤维素
centrifugal force　地心引力
chain initiation　链引发
chain transfer agent　链转移剂
chain-growth　链增长
chain-reaction　链式反应
chain-transfer agent　链转移剂
chemical bond　化学键
chemical combination　化学组成
chemical tubing　化工管道
chemically inactive　化学惰性
chemically polymerized　化学聚合
chopped fiber　碎纤维
chrome plated roll　镀铬辊
closed cell structure　闭孔结构
closed mold　合模
cloth fabric　（布）织物
coated sheet　涂布片材
cohesive energy density　内聚能密度
cold drawing　冷拉伸
cold slug well　冷料井
collapsing plate　压扁[平]板，夹膜板，人字板
color concentrate　色母料（粒）
color possibility　着色性
combination of properties　综合性能
complex initiator system　复合引发体系
composed of　由……组成的
compression mold　压模
compression molding　压制模塑
compression process　压塑加工
compression section　压缩段
compressive load　压缩负荷
computer-aided approach　计算机辅助方法
concentrate on　将……集中于……
condensation polymer　缩聚物
condensation polymer　缩（合）聚（合）物
condensation polymerization process　缩聚方法，缩聚过程
condensation reaction　缩合反应
configurational state　构型状态
contact cement　接触胶合剂
control console　控制[操纵]台
conveyor system　输送器，传送机系统
cooling channel　冷却水通道
cooling roll　冷却辊
cooling trough　冷却槽
cooperative wriggling　共同蠕动
co-ordination compound　络合配位化合物
coordination polymerization　[高分子]配位聚合
copolymerization theory　共聚理论
correspond to　相当于
cosmetic cases　化妆盒
counter-rotating twin-screw extruder　反向旋转双螺杆挤出机
covalent bonding　共价键
covalent crosslink　共价键交联
crankshaft motions　曲轴运动
crash pad　防震垫
creep compliance　蠕变柔量
creep test　蠕变测试
cross section　横截面
cross-contamination　交叉污染
crossed polarizer　正交偏光镜
cross-linked network　交联网状
cross-linking agent　交联剂
cross-sectional area　断面面积
crystal clear material　晶状透明材料
crystal lattice　晶格，晶体点阵

crystalline polymer　结晶聚合物
cube shaped　方形的
curing agent　固化剂,硫化剂
dash panel　(汽车的)仪表板
decorative laminates　装饰层压板
degree of crystallinity　结晶度
degree of polymerization　聚合度
derived from　从……中产生
diameter gauge　直径测量
die land　口模区
dielectric spectroscopy　介电谱
dielectric thermal analysis　热介电分析
differential scanning calorimetry　差示扫描量热法
differential thermal analysis　示差热分析
difunctionality　二官能性
dilute solution viscosity　稀溶液黏度
dimensional stability　尺寸稳定性
dimensional tolerances　尺寸公差
dip coating　蘸涂
dipole moment　偶极矩
dispersion force　分散力
dispersive mixing　分散混炼
dissociation energy　离解能
distributive mixing　分布混炼
distributor roll　分散辊,(油墨)匀布辊
double bond　双键
down to　降到
drag flow　黏性流,阻曳流
draw ratio　拉伸比
dreamed of　梦想
drilling muds　钻井泥浆
drive shank　驱动段
driven punch roll　传动轧辊
dry spinning　干式纺丝,(亚麻)干纺
Du. Pont Company　杜邦公司
dump extruder　卸料挤出机
eccentricity gauge　偏心测量
eclipsed position　重叠的位置
ejector mechanism　顶出机构
ejector pins　顶杆
elastic modulus　弹性模数
elastic recovery　弹性恢复
elastic solid　弹性固体
electrical insulation　电绝缘
electrical panel　配电板罩
electronegative chlorine atom　负电性的氯原子
electrons cluster　电子簇,电子雾,电子云
ellipsoidal shape　椭圆形
embossed roll　压花辊
emerge from　从……露出,从……浮现,来自,产生于
empirical formula　经验化学式
emulsion polymerization　乳液聚合
emulsion polymerized polybutadiene　乳化聚丁二烯
end group　端基
energy dissipation　能量损耗
energy resource　能源
engineering material　工程材料
enthalpy change　热含量变化
epoxy resins　环氧树脂
equivalent masses　等效质量
esterification reaction　酯化反应
ethylene dichloride　二氯化乙烯
ethylene glycol　乙二醇
ethylene-vinyl acetate　乙烯-醋酸乙烯酯
exhaust system　排气系统
expandable bead　可发珠粒料
extent of crosslinking　交联度
extruder screw　挤出螺杆
extruder screws　挤出机螺杆
extrusion rate　挤出速率
fairly concentrated polymer solution　聚合物浓溶液
fatigue resistance　耐疲劳
fatty acid　脂肪酸
feed hopper　加料斗

feed section 加料段
female mold 阴模,下半模,凹模
find use in 应用于
finished article 成品
finishing equipment 压光机
finishing roll 光辊
fire resistance 耐火
flat sheet (平)片材
flex cracking 屈挠龟裂
flex life 挠曲寿命
flexible sheet 软片
flexural strength 挠曲强度
flow work 流动功
fluidized bed coating 流化床涂布
fluorocarbon polymer 氟碳高聚物
foamed-in-place 现场发泡
foaming action 发泡作用
foaming process 发泡过程
formed plastics 泡沫塑料
free blowing (热成型)自由吹胀成型(即无模吹气成型)
free radicals 自由基
free surface 自由表面
fringed-micelle model 缨状微束模型
frozen out 冻结
functional group 官能团
fundamental bases 基本依据
fuse block 保险丝装置,熔丝盒
fusion cycle 塑化周期
gas chromatography 气相色谱分析
Gaussian chain 高斯链
gear-like roller 齿轮状辊
gegen or counter ion 反离子,抗衡离子
generalized flowcharts 工艺流程图
giant molecule 大分子
give rise to 引起,导致
glass mat 玻璃毡片
glass transition temperature 玻璃化转变温度
glassy amorphous state 玻璃无定形状态
glassy state 玻璃态
glue gun 黏合剂用喷枪,喷胶器,热融枪
graft copolymer 接枝共聚物
granular polymer 粒状聚合物
grinding wheel 砂轮
guide roller 导向辊
gusset bar (吹膜机)边折板
gutta-percha 杜仲胶
hard liner 硬衬套
haul-off(= take off) 引出,引出装置
head-to-tail configuration 头尾构型
heat distortion point 热扭变点
heat resistance 耐热性
heater bands 加热圈,电热圈(环)
heuristic principle 启发性原则
hildebrand function (theory) 希耳德布兰特函数(理论)
hollow pellet 空心粒料
hook up 连接
hot material feed 热材料供料
hot-stage (显微镜)热台
hydraulic press 液压机
hydraulic pressure 液压
hydrogen bond 氢键
hydrogen chloride 氯化氢
hydrolyzed collagen 水解胶原蛋白
hydroxy acid 羟基酸
hydroxyl group 羟基
ice cube (加入饮料用的)小方冰块
identical to 与……一致
identification card 身份证
imitation marble 仿大理石
impact strength 冲击强度
in accordance 与…一致
in batches 分批地,成批地,批量地
in combination 结合
in competition with 与……竞争
in conjunction with 在……同时
in equilibrium 平衡

in increment　分批量,分期地
in moldable form　处于可模塑的形式
in terms of　用术语
in the business of　从事
in the case of　就……来说,关于
individual lamellae　个别晶片
inert solvent　非极性溶剂
Infrared spectroscopy　红外光谱分析
injection molding machine　注塑机
input capstan　送料(输入)绞盘
insulation recipe　绝缘配方
integral part　组成部分
integral unit　整体部件
interconnected cell　联泡孔
interfacial phase　界面相
internal mixer　密炼机
intrinsic viscosity　特性黏度
irregular cross section　异型材
irreversible process　不可逆过程
irreversible reaction　不可逆反应
isotactic polymer　等规聚合物,[高分子]　全同立构聚合物,全同聚合物
isotactic polypropylene　全同立构聚丙烯
it might appear　看来,似乎
job shop　加工车间
keep in mind　记住,放在心里
knock-out bars　顶出杆
know-how　专业知识
kraft paper　牛皮纸
laminar flow　层流
laminated sheet　层压板
large particle size diluent　粗粒增容剂
latex paint　乳胶涂料
leftover proton　过剩的质子
length-diameter ratio　长径比
length-to-diameter ratio　长径比
liable to　易于……的,有……倾向的
life jacket　救生衣
linear conformation　线形构象
liquid-crystalline　液晶
living organism　生命有机体
load bearing capacity　承载能力
loading tank　加料槽
long-chain　长链
loss modulus　损耗模量
low-angle X-ray diffraction　小角 X 射线衍射
lower critical solution temperature(LCST)　最低临界溶解(相容)温度
lyotropic liquid-crystal　溶致液晶
machined dies　机头,模头
macromolecular hypothesis　大分子假说
magnesium oxide　氧化镁
magnitude of stress or strain　应力或应变的大小
main-chain bond　主链键
make sth. ideal for　使……适合于
male mold　阳模
mandrel programmer　芯棒控制装置
mass spectrometry　质谱分析法
matched die　对模
matched-mold forming　对模成型
mathematical treatment　精确处理
Maxwell element　麦克斯韦单元
mechanical entanglement　机械缠结
mechanical model　力学模型
mechanical properties　力学性能,机械性能
mechano-chemistry　力化学
melamine-formaldehyde　三聚氰胺-甲醛树脂
melt fracture　熔体破裂
melt spinning　熔融纺丝
metering pump　计量泵
metering section　计量段
methacrylic acid　甲基丙烯酸,丙烯酸
methyl group　甲基
methyl methacrylate in benzene　甲基丙烯酸苯
mold cavity　模具型腔

mold halves 模具组分
molded block 模体,模块
molded part 模塑制件
molding cycle 模塑周期
molecular dipole 分子偶极子
molecular weight 相对分子质量
more than 不只,超过
more-or-less 或多或少,大体上,大约,左右
mounting plate 固定板
moving platen 动模板
multiple dies 复合机头
multiple piece molds 多腔模具
naked eye 肉眼
natural fiber 天然纤维
natural polymer 天然高分子
natural rubber 天然橡胶
natural state 自然状态
necking down 成颈,缩颈
neoprene rubber 氯丁橡胶
nitrile rubber 丁腈橡胶
nonpolar solvents 非极性溶剂
nonpolymeric solutes 非聚合物溶质
non-uniform 非均匀性
no-stick cookware 不粘炊具
null-balance 零位平衡
number-average molecular weight 数均相对分子质量
off-white 纯白色的,米色的
on the order of 大约,数量级为
one cycle (注射的)一个周期
one-dimensional structure 一维结构
open cell structure 开孔结构
open cellular structure 开孔结构,开放的网格结构
order of magnitude 数量级
ordinary low molecular weight compounds 普通低分子化合物
organic acid 有机酸
organic base 有机碱
organic peroxide 有机过氧化物
organic reaction 有机反应
organic substance 有机物
owing to 因为,由于,归因于
packaging industry 包装工业
parting line 分模线,合模线
pass through 流过,通过
patten roll 压花辊(模型辊)
pay-off drum 送料卷盘
pendant group 侧基
pendant group 侧基
peptising agent 塑解剂
phase transformation 相转变
phenol-formaldehyde 苯酚-甲醛
physical ageing 物理老化
physical foaming 物理发泡
picture frame 画框
pinch off 夹断,压紧
piston ring 活塞环
plotting temperature 策划温度
pneumatic press 气动压力机,气压机
pointing upwards 指出向上
polar group 极性基团
polar secondary force 极性次价力
polar solvent 极性溶剂
polar solvents 极性溶剂
poly ethyl acrylate 聚丙烯酸乙酯
poly(vinyl alcohol) 聚(乙烯醇)
poly(vinyl chloride) 聚氯乙烯
polyethylene terephthalate 聚对苯二甲酸乙二酯
polymer chain 聚合物链
polymer physics 高分子物理学
polymer single crystal 聚合物单晶
polymer-based composites 聚合物基复合材料
polymethyl methacrylate 聚甲基丙烯酸甲酯
polyphenylene oxide 聚氧化乙烯
polyvinyl acetate 聚乙烯乙酸酯

polyvinyl alcohol　聚乙烯醇
polyvinyl chloride　聚氯乙烯
power drill　机械钻
power input　功率输入
precipitating agent　沉淀剂
precision-engineered components　工程零件
premixed molding　预混模制
pressure forming　压力成型法
pressure-sensitive tape　压敏胶黏带
primary bond　主价键
primary covalent bond　主价键
profile shape　型材
propagation reaction　链增长反应
proportional to　与……成比例
protective coating　保护涂层
prototype part　原型零件
pull rolls　引出辊
pull-off roll　拖出辊
push-back pin　回程杆
quantitative treatment　定量处理
radiation curing　辐射硫化
random copolymer　无规共聚物
raw material　原材料
ready for　预备好
ready-to-fill adj.　现成的，做好的
reciprocating screw　往复式螺杆
reclaim rubber　再生橡胶
reclaimed rubber　再生橡胶
regardless of　不管，不顾
rejected product　次品，废品
relaxation modulus　松弛模量
relaxation process　弛豫过程
rely upon(rely on)　依靠
repeat unit　重复单元
repeating unit　重复单元
residence time　停留时间
residence time distribution　停留(阻滞)时间分布
resilient floor　弹性地板
resist tear　耐撕裂
resistance to crushing　抗压裂性
restore force　回复力
restrictor bar　限流块
retainer plate　垫板
rigid pipe　硬管
rolling nip　辊距
rubber band　橡皮圈
rubber cement　橡胶胶水
rubber covered　涂胶
rubber plateau region　橡胶平台区
rule-of-thumb　经验法则
saturated hydrocarbon　饱和烃
schematic diagram　原理图，示意图
science of synthetic　合成材料学科
scratch resistance　抗刮伤
screen pack　(挤出机的)滤网叠
screw diameter　螺杆直径
screw flight　螺杆的螺纹
screw-ram　螺杆-射料杆式
secondary bond　次合键，次价力键
second-order phase transition　二级相变点
self-extinguishing　自熄性
self-purging　自行清洗
semireinforcing fillers　半补强填料
separate…from　把……从……中分开
shared electrons　共价电子，共用电子对
shear deformation　剪切变形
shear rate　剪切速率
shear strength　剪切强度
shear-thickening　剪切增稠
shear-thinning　剪切变稀
sheet material　片材
side branch　支链，分枝
side reaction　副反应
silly putty　弹性橡皮泥
simple shearing flow　简单剪切流
simulated leather　仿皮革
single bond　单键

sintering powdered 粉料烧结的
size exclusion chromatography 体积排阻色谱
slab stock 泡沫塑料，块料
snap-back forming （热成型）快速吸塑成型，骤缩成型
snap-on lid 带按扣的盖子
sodium chloride crystal 氯化钠晶体
softening agent 软化剂
softening point 软化点，软化温度
solvent resistance 耐溶剂性
spark tester 电火花实验（检测）
special purpose rubber 特种橡胶
specific volume 比体积
spinning block 旋压模
split out 分裂，排出
split up into 分割成……
sponge-like material 类海绵材料
spring loaded 弹簧加载
sprue bushing 浇口套
stainless steel 不锈钢
static electricity 静电
stationary platen 定模板
steady-state 不变的状态，平衡状态，稳定状态
stem from 由……发生（产生，引起）
step-wise 逐步
steric（geometric）effect 立体（几何）效应
steric repulsion 位阻排斥力
storage modulus 储能模量
straight pressure forming 直接压力成型
strained ring 张力环
stress concentrator 应力集中点
stress crack 应力开裂
stress relaxation 应力松弛
strong acid 强酸
strong Leuis acid catalyst 强路易斯酸催化剂
structural application 结构上的应用
styrene-acrylonitrile 苯乙烯丙烯腈
sun visor 遮光板
support pillar 支承柱
surface attachment 表面吸附
surface finishes 表面加工，表面修饰
surface gloss 表面光泽度
swept-out 被卷走
synthesize rubber 合成橡胶
synthetic fiber 合成纤维
synthetic rubber 合成橡胶
tailor-make 特制的，适合的
take sth. for granted 认为……是理所当然
take-off roll 牵引辊
take-off system 牵引系统
take-off unit 卷曲装置
takeup velocity 引出速度，卷取速度
tapered guide 锥形导线器
technological approaches 技术途径
tension control 张力控制
tension controller 张力控制器
termination reaction 终止反应
that is to say 即，也就是说，换句话说，更确切地说
the crystalline bonding forces 结晶黏结力
the gel effect 凝胶效应
the Gibbs free energy 吉布斯自由能
the isochronal high-temperature process 等时的高温过程
the order of 顺序
the reverse order chronologically 倒序顺序
thermal energy 热能
thermal expansion 热膨胀
thermal expansion coefficient 热膨胀系数
thermal insulation 绝热
thermal mechanical analysis 热机械分析

thermal motion　热运动
thermal optical analysis　热光学分析
thermal shock resistance　耐热骤变性
thermodynamics and statistics　热力学与统计学
thermoplastic polymer　热塑性聚合物
thermosetting plastics　热固性塑料
thin-wall　薄壁的
three-dimensional crystal lattice　三维晶格
three dimensional network　三维网状结构
three dimensions　三维空间
time-dependent　有时间依赖性
time-temperature superposition　时-温等效叠加
tissue paper　棉纸,薄纸
to the extent　在……程度上
torsion pendulum　扭转摆
torsional angle　扭转角
transfer molding　传递模塑
translational motion　平移运动
trial-and-error　反复实验,尝试法,逐步逼近法,试错法
tumbling-type　滚动式
twist into thread　捻成一根
two-cavity injection mold　双腔注模
two-dimensional lattice model　二维格子模型
two-roll mill　双辊塑炼机,双辊开炼机
uniaxial tensile　单轴拉伸
unsaturated polyester　不饱和聚酯
upper critical solution temperature (UCST)　最高临界溶解(相容)温度
upper die jaw　上模唇
vacuum forming　真空成型
van der Waals　范德瓦尔斯(范德华)
velocity gradient　速度梯度
vent hole　排气孔,通风孔
vented extruders　排气式挤出机
vinyl acetylene　乙烯基乙炔
vinyl resin　乙烯基树脂
viscometric flow　黏性流动
viscous fluid　黏性流体
vulcanizing agents　硫化剂
water resistance　耐水性
water-soluble　水溶性的
wavy surface　波纹表面
wear resistance　耐磨性
weather resistance　耐候性
Weissenberg effect　韦森堡效应
wet process　湿法加工,湿法
wet spinning　湿纺,湿法纺丝
wetting out　浸润
white lead　铅白
wider and thinner sheet　片材变宽变薄
wind-up　收卷,收卷机,结局,结尾
windup drum　卷取卷盘
with respect to　关于
within the range of　在……范围内
x-ray diffraction pattern　X 射线衍射图
zero-entropyproduction　零熵产生
zero-shear viscosity　零切黏度

Primary References
主要参考文献

[1] 揣成智. 聚合物科学与工程导论[M]. 北京:中国轻工业出版社, 2010.

[2] 揣成智. 高分子材料工程专业英语[M]. 北京:中国轻工业出版社, 1999.

[3] Ronald J. Baird, Industrial Plastics[M]. South Holland, III: The Goodheart-Willcox Co., Inc. Publishers, 1971.

[4] Bajal Mahendra D. ed, Plastics polymer science and technology[M]. New York: Wolley, 1982.

[5] Evans Cotin Wynne, Practial rubber compounding and processing[M]. London: Applied Science Publishers, 1981.

[6] Donald V. Rosato, Dominick V. Rosato, Blow molding handbook[M]. New York: Hanser/Gardner Publications, Inc., Cincinnati, 1989.

[7] Dominick Rosato, Donald Rosato, Plastics Engineered Product Design[M]. Oxford: Elsevier Advanced Technology Publisher, 2003.

[8] Ulf W. Gedde, Polymer Physics[M]. London: Chapman &Hall Publisher, 1995.

[9] Utracki L. A. Polymer Blends Handbook[M]. Netherlands: Kluwer Academic Publisher, 1999.

[10] Lawrence E. Nielsen, Probert F. Landel[M]. New York: Marcel Dekker, Inc., 1994.

[11] Stephen L. Rosen, Fundamental principles of polymeric materials[M]. New York: A Wiley-Interscience Publication, 1993.

[12] Dominick Rosato, Donald Rosato, Plastics Engineered Product Design[M]. Oxford: Elsevier Advanced Technology Publisher, 2003.